Polysaccharides in Advanced Drug Delivery

Polysaccharides in Advanced Drug Delivery

Akhilesh Vikram Singh, PhD
UGC-DSKPDF, Department of Materials Engineering,
Indian institute of Science (IISc),
Bangalore, Karnataka, India

Bang-Jing Li, PhD
Professor, Chengdu Institute of Biology,
Chinese Academy of Sciences (CAS),
Chengdu, Sichuan, China

CRC Press is an imprint of the
Taylor & Francis Group, an **informa** business

First published 2020
by CRC Press
2 Park Square, Milton Park, Abingdon, Oxon, OX14 4RN

and by CRC Press
6000 Broken Sound Parkway NW, Suite 300, Boca Raton, FL 33487-2742

CRC Press is an imprint of Informa UK Limited

British Library Cataloguing-in-Publication Data
A catalogue record for this book is available from the British Library

Library of Congress Cataloging-in-Publication Data
A catalog record has been requested

ISBN: 978-0-367-46220-8 (hbk)
ISBN: 978-1-003-03263-2 (ebk)

Preface

In the past few decades, the role of polymers, i.e. natural as well as synthetic polymer has been increased in the field of medical research, particularly in the disease diagnostics and advanced drug delivery system. Despite several advantages such as better biocompatibility, cost-effectiveness and easy availability, natural polymers were less explored commercially compared to its synthetic equivalent.

The basis of this book is to expose the reader to the upcoming area of natural polymers and its potential application in the field of advanced drug delivery. The chapters are written by some of the most esteemed scientists in the field of material science and drug delivery. All the chapters are written to emphasize physicochemical properties of respective polysaccharides and their application in medical science.

This book, intended for postgraduate student instructions as well as it can also be used as a reference standard for research students, faculty and industry people. This book gives a full insight of different natural polymers and their application in the field of advanced drug delivery. Each chapter is written so as to give a broad overview of a topic and concluded with most recent up to date information on their utilization in the field of pharmaceutical research. By introducing this utmost needed book, we are hopeful that we will inspire the next generation of researchers to make better contributions to the emerging and dynamically changing field of advanced drug delivery.

We would like to thank all the authors for their valuable contributions and making it a successful journey. We would also like to thank BSP Books Pvt. Ltd. Hyderabad, India for their help in organizing the chapters, editing, and providing assistance in general.

- Authors

Contents

1 Polysaccharide Carriers for Induction and Evaluation of Tissue Regeneration and Drug Delivery

Jun-ichiro Jo[1] and Yasuhiko Tabata[2]*

[1]Diagnostic Imaging Program, Molecular Imaging Center, National Institute of Radiological Sciences, 4-9-1 Anagawa, Inage, Chiba 263-8555, Japan.

[2]Department of Biomaterials, Institute for Frontier Medical Sciences, Kyoto University, 53 Kawara-cho Shogoin, Sakyo-ku Kyoto 606-8507, Japan.

1.1 Introduction

Induction of tissue regeneration has been highly expected as a new field of medical treatment covering or compensating two advanced medical therapies; reconstruction and organ transplantation therapies. The basic idea of tissue regeneration induction is to regenerate or repair the injured or lost tissues and substitute organ functions based on the natural self-healing potential of patients themselves. With the recent rapid development of cell biology, it has been possible to make use of various progenitor and stem cells with high potentials of proliferation and differentiation for cell-based induction of tissue regeneration. It has been demonstrated that cells themselves have good therapeutic potentials in terms of their inherent targetability to the site injured or biological properties. However, the therapeutic efficacy of cells transplanted is not always as high as expected, which is one of the largest problems in cell therapy. This is because the survival rate of cells transplanted is low, and consequently the biological functions of cells are not always expected in the body. To tackle the problem, it is indispensable to develop materials, technologies, and methodologies to provide the cells a local environment where the survival and biological functions of cells transplanted can be maintained or enhanced. On the other hand, the technologies and methodologies to evaluate the extent and process of tissue regeneration still depend on the conventional diagnostic methods such as histological, biochemical, and morphological examinations. Clinical availability of these examinations is limited in terms of their invasiveness and

*** *Corresponding Author***

reliability. Therefore, it is also necessary to develop new systems for the non-invasive evaluation of tissue regeneration. On the basis of these requirements, many types of biomaterials (polymers, metals, ceramics, and their composites) as cell scaffolds and drug delivery carriers have been designed and created to achieve the effective induction and evaluation of tissue regeneration.

Polysaccharide is a monosaccharide-repeated biopolymer with various glycosidic bonds and composes a part of living body such as wall of plant cells, exoskeleton of arthropods, connective tissue or cell surface as well as serves as energy yielding fuel. Most of polysaccharides have long histories of the medical, pharmaceutical, and food applications, and their biosafety and bioavailability have been proven based on the practical usage. Since polysaccharide has reactive groups, such as hydroxyl, carboxyl or amino groups, the chemical modification is easy to give it chemical, physical, and biological properties and to form several types of structures. Taken together, polysaccharide is expected to be the one of the most feasible biomaterials for the effective induction and evaluation of tissue regeneration. In this chapter, the current status in the induction and evaluation of tissue regeneration based on polysaccharide as a cell scaffold and drug delivery carrier is overviewed. Additionally, recent researches on the manipulation and tracing of stem cells and molecular imaging of tissue regeneration with a nano-sized polysaccharide construct are introduced.

1.2 Role of Biomaterials for Induction of Tissue Regeneration

It is well recognized that cells are present in the living tissue interacting with the extracellular matrix (ECM) of a natural scaffold for their proliferation, differentiation, and morphogenesis. When the body tissue is largely lost, the ECM itself also disappears. In such a case, only by supplying cells to the defect, we cannot always expect the natural induction of tissue regeneration. There are two approaches to induce the effective tissue regeneration based on the concept of tissue engineering[1], which provides cells transplanted with a suitable environment where the cells maintain their biological functions such as survival, proliferation, and differentiation (Figure 1.1). One is to provide a temporary scaffold and bioactive substance (bio-signaling protein or the related gene) to the defect for the proliferation and differentiation of cells. The other is to transplant cells with the biological function manipulated by the cellular internalization of bioactive substance. It is indispensable to design the practical biomaterials that support these approaches. Following sections describe several key points for biomaterials design to make these approaches effective.

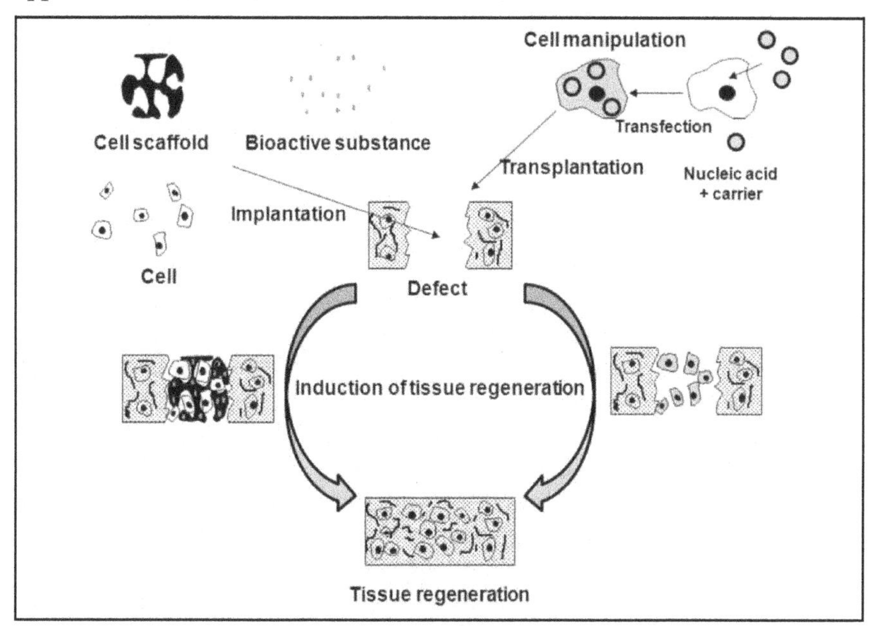

FIGURE 1.1 Tissue engineering approaches for induction of tissue regeneration.

Cell scaffold materials should be safe contacting with cells transplated and have an adequate mechanical strength during the process of tissue regeneration while they should disappear after the tissue regeneration. Therefore, the cell scaffold materials should be non-toxic, biocompatible, and biodegradable. The ECM is composed of structural proteins (collagen or elastin etc.), adhesion proteins (fibronectin, laminin, vitronectin, etc.), and several polysaccharides (glycosaminoglycans; hyaluronic acid or sulfated polysaccharides). In this context, these kinds of materials are mainly used for preparation of scaffolds to mimic a native extracellular environment. Since water-soluble polysaccharides are rapidly diffused to disappear from the site placed, they should be insolubilized to use as a scaffold.

Insolubilization is achieved by

(i) The ionic interaction with low-molecular-weight counter ion,
(ii) The formation of polyion complex with polyelectrolytes having an opposite charge,

(iii) The formation of interpenetrating polymer network with nonionic polymer,
(iv) The acquisition of temperature sensitivity by modification with a hydrophilic or hydrophobic residue, and
(v) The formation of chemical bonding by a crosslinking reagent or radiation irradiation.

Three-dimensional and sponge scaffolds with a pore interconnectivity should be prepared to provide cells with the sufficient surface area for their initial attachment and the adequate supply of nutrient and metabolite. Processing methods to provide the porous structure in the scaffolds include particle reaching, freeze drying, phase separation, fiber meshes, fiber bonding, melt processing, batch forming, electrospinning, and rapid prototyping[2]. Different shapes of three-dimensional porous scaffolds, such as granule, sheet, fiber, fiber mesh, and non-woven fabric, can be obtained by various processing methods. In addition to the porous scaffold, the preparation of injectable scaffolds is another feasible approach for the induction of tissue regeneration. The advantage of injectable scaffolds is to allow to avoid the invasive surgical implantation and to easily fill the individual and irregular shape of defect. Injectable scaffolds can be divided into two types; microparticle and *in situ* forming scaffolds. Microparticle scaffolds are prepared by the methods of precipitation, simple or complex coacervation, spray drying or suspension, emulsion, and dispersion polymerization/crosslinking. *In situ* forming scaffolds are prepared via thermoplastic pastes, *in situ* polymerizing/crosslinking, *in situ* precipitation, and gelation under an environment-based stimulation. Cell scaffolds can be prepared from not only one kind of biomaterial but also a biomaterial-biomaterial composite. By making use of the composite, cell scaffolds with an easy processing of preparation or an improved biological or mechanical property can be obtained.

Combination of scaffold with a bioactive substance efficiently induces tissue regeneration. Since the bioactive substance rapidly diffuses from the injected site and is enzymatically digested or deactivated, it is necessary to develop the technologies and methodologies to effectively deliver the bioactive substance to the site to be regenerated based on the concept of drug delivery system (DDS). DDS is a technology which allows a drug to act at the right time the right site of action at the appropriate concentration. The objectives of DDS include the controlled release, the life-time prolongation, the accelerated permeation and absorption, and the targeting of drug (Figure 1.2). Various biomaterials have been extensively used to achieve each DDS objective. Since most of scaffold biomaterials are biodegradable, they are expected to be applied for the controlled release of bioactive substances by their incorporation into the scaffold. Incorporation of bioactive substances is carried out during or after processing of scaffold preparation, which depends on its sizes. Incorporation into a biomaterial scaffold enables the bioactive substance to protect from the enzymatic attack and gradually release in the site to be regenerated. Release profile of bioactive substance is governed not only by the degradability or structure change of scaffold biomaterials, but also by the interaction strength with the scaffold biomaterial chain. Therefore, it is necessary to design and construct scaffold biomaterial based on these aspects for the ideal controlled release of bioactive substances.

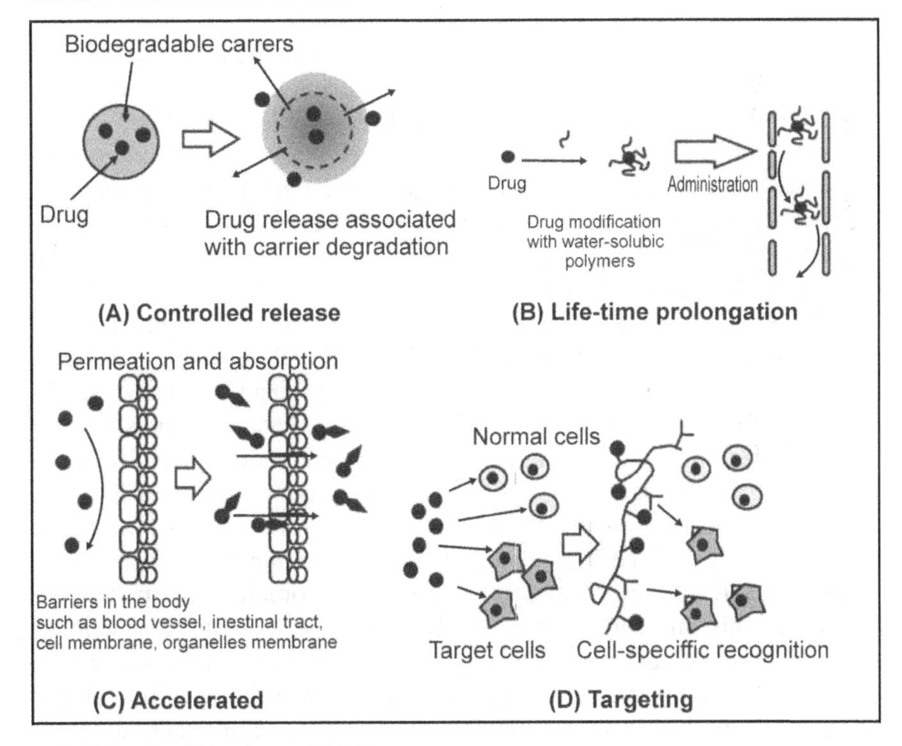

FIGURE 1.2 Objectives of DDS.

Transplantation of cells artificially manipulated to obtain biological functions, such as the survival, proliferation, differentiation, and therapeutic abilities, is expected to promote the induction of tissue regeneration compared with the original cells alone. Generally, nucleic acids, such as plasmid DNAs, antisence oligonucleotides or small interfering RNAs (siRNA), are mainly used as bioactive substances for the manipulation of cell functions. The manipulation of cell function is fundamentally achieved by the transfection procedure of the related nucleic acids for the cells. Therefore, it is necessary for successful cell manipulation to develop materials, technologies, and methodologies for safe and efficient transfection of nucleic acid. The transfection system is generally divided into two categories in terms of carrier materials of nucleic acids: viral and non-viral systems. For the viral system, the carrier of retrovirus, lentivirus, adenovirus, and adeno-associated virus, has been used to be potentially efficient, although there are several issues to be resolved for the clinical applications, such as the antigenicity and toxicity of virus itself or the possibility of disease transfection. Therefore, efficient technologies and methodologies of transfection without using the virus are highly expected. The non-viral system has several advantages in terms of its safety and no limitation in the molecular size of nucleic acid applied. However, the low transfection efficiency is one of the major drawbacks for the research and therapeutic applications. In this circumstance, various experimental trials have been performed to develop the physicochemical and biological properties of non-viral carriers. Nucleic acid in the naked form is a polyanion of phosphate group-repeated chain and has an expanded molecular structure due to the intermolecular repulsion force of negative charge at the physiological pH. Therefore, it is well recognized that the nucleic acid cannot interact with the cell membrane negatively charged due to the electrostatic repulsion, and consequently hardly be internalized and subsequently transfected. On the basis of these findings, the non-viral carriers of cationized polymers[3-5] and cationized liposomes[6-8] have been developed to allow nucleic acid to effectively internalize into cells for transfection (Figure 1.3). The cationized materials enable the nucleic acid to form a complex with a molecular size and surface charge which are suitable for the cellular internalization and consequently enhance the complex internalization. However, this method is only to increase the non-specific cellular internalization of nucleic acid through the simple electrostatic interaction between the complex and cell surface. To enhance the specific cellular internalization of complex by making use of biological interactions, there have been several research trials with ligands specific for cell surface receptors, such as folate, transferrin, mono- or oligosaccharides, peptides, and proteins. Modification of the ligand will enable nucleic acid to internalize into cells in a cell-specific manner. It is also important for effective biological expression of nucleic acid to consider the intracellular trafficking of nucleic acid-carrier complexes or their stability in the cell.

In the case of normal intracellular trafficking, the nucleic acid-carrier complex internalized via an endocytosis pathway is carried into the endosomal compartment, followed by the lysosomal degradation. Therefore, the carrier should be molecularly designed to allow the nucleic acid-carrier complex internalized to effectively escape from the endosomal compartment into the cytoplasm one. For example, when covalently linked with a peptide capable to disrupt the lysosomal membrane under an acidic condition where lysosomal enzymes biologically function, a carrier effectively enhanced the expression level of nucleic acid[9]. It is reported that the carrier covalently linked with functional groups having a buffering capacity to accelerate the endosomal escape, so-called "proton sponge effect", enhanced the expression level of nucleic acid[10]. Furthermore, the carrier covalently linked with a peptide of nuclear localization signal (NLS) enabled a nucleic acid to positively deliver to the cellular nucleus[11]. There have been several researches to combine nucleic acid-carrier complexes with physical stimuli, such as pressure, electricity, ultrasound, magnetism, and light, to enhance or regulate the level and pattern of nucleic acid expression.

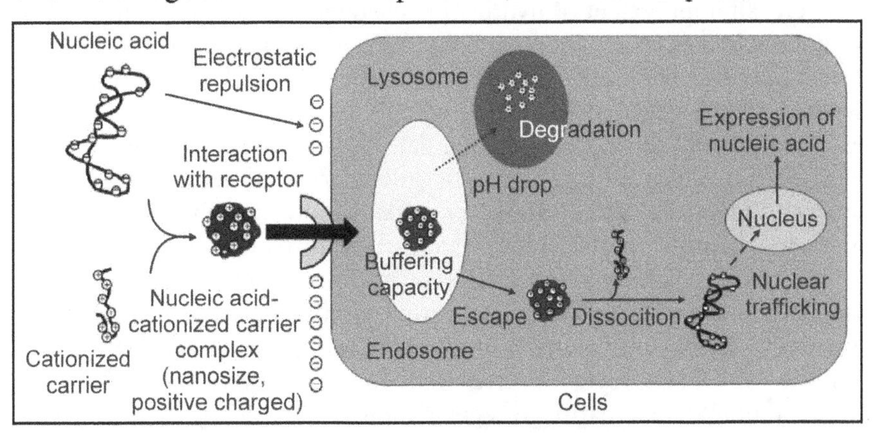

FIGURE 1.3 Internalization and intracellular trafficking of nucleic acid complexed with a non-viral cationized carrier for the expression.

1.2.1 Polysaccharide as a Cell Scaffold to Induce Tissue Regeneration

As described above, many kinds of biomaterials have been used for the preparation of scaffolds. Polysaccharide is largely non-toxic, biocompatible, biodegradable, and often degraded by specific enzymes in the body. In addition, the bioadhesive property and low cost for acquisition are advantageous over other biomaterials. It is well known that glycosaminoglycans (GAG) are one component of ECM. The GAG possess biological characteristics of the binding and modulation of growth factors and cytokines of bioactive substances, the inhibition of proteases' function, and the involvement in adhesion, migration, proliferation, and differentiation of cells[12]. Therefore, making use of polysaccharide with chemical structures similar to that of GAG is quite advantageous to prepare biologically functional cell scaffolds. Furthermore, several polysaccharides have properties of self-insolubilization or *in situ* crosslinking, which have been practically used as injectable scaffold systems. There have been many types of polysaccharides to use as cell scaffolds to induce regeneration of various tissues. The following describes the representative polysaccharides as cell scaffold for induction of tissue regeneration.

Alginate (Figure 1.4A) is an anionized polymer linked with β-D-mannuronic acid and α-L-guluronic acid units. Alginate is insolubilized to form a hydrogel in aqueous solution by the ionic interaction of α-L-guluronic acid units with divalent cations such as calcium or barium ions. The crosslinking extent of hydrogel is changed by the composition of α-L-guluronic acid in the alginate. Since the hydogel is formed under a mild condition, alginate has been widely used as cell scaffolds for induction of various tissues regeneration. The hydrogel or beads of alginate scaffolds can be prepared by applying the alginate aqueous solution containing the desired bioactive substance or cells into a solution containing divalent cations. Other insolubilization methods to prepare the cell scaffold include chemical crosslinking[13], complex formation with cationized polymer[14], and composite formation[15]. Alginate has often been mainly applied for the induction of cartilage regeneration because the structure is similar to the GAG of chondrocytes' ECM and that the three-dimensional culture of chondrocytes is required for the induction of cartilage regeneration[16]. The cartilage regeneration is induced by alginate beads, discs, hydrogels, and sponges encapsulating chondrocytes or progenitor/ stem cells combined with several bioactive substances[17-20]. Other researches on the regeneration induction of various tissues, such as vascular, peripheral nerve, intervertebral disc, liver, and pancreas, with alginate scaffold have been also reported[13,21-24].

Chitosan (Figure 1.4B) is a cationized polymer composed of N-acetyl-D-glucosamine and D-glucosamine. Chitosan is enzymatically degraded by the lysozyme. The water-solubility and biodegradability are altered by the relative proportions of N-acetyl-D-glucosamine and D-glucosamine in the chitosan molecule and the solution pH. To use as scaffolds, chitosan is crosslinked mainly by the crosslinking reagent or formation of polyion complex. Different shapes of chitosan-based scaffold have been prepared, for example, granules, fiber, tube, sponge or microsphere. Since the structure of chitosan is also similar to the GAG of chondrocytes ECM[12], chitosan microsphere scaffold have been used for cartilage regeneration[25]. In addition, it has been reported that chitosan has a strong tissue-adhesive property and enhances blood coagulation[26]. In this context, chitosan hydrogel scaffold has been used as a wound dressing[27]. Regeneration of other tissues, such as bone, vascular, and peripheral nerve, has been reported by using the scaffold with or without the incorporation of the related bioactive substance[28-30].

Hyaluronic acid (Figure 1.4C) is an anionized polymer composed of α-1,4-D-glucuronic acid and β-1,3-N-acetyl-D-glucosamine. Since the hyaluronic acid is highly water-soluble, the chemical modifications are required to insolubilize. Hyaluronic acid modified with benzyl ester groups is called HYAFF®, which is widely used for a cell scaffold material[31]. Various shapes of HYAFF® have been produced and applied for the regeneration induction of cartilage, vascular, and adipose tissues[32-34]. Other researches have been also reported on the regeneration induction of spinal cord[35] and skin[36] tissues with hyaluronic acid-based scaffold.

In other polysaccharides, scaffolds prepared by agarose (Figure 1.4D)[37], cellulose (Figure 1.4E)[38,39], chondroitin sulfate (Figure 1.4F)[40-43], dextran (Figure 1.4G)[44,45], gellan gum (Figure 1.4H)[46] or starch (Figure 1.4I)[47,48], have been used for induction of tissue regeneration by making use of their individual properties.

1.2.2 Polysaccharide as a Transfection Carrier to Induce Tissue Regeneration

Among the non-viral materials applicable for the transfection carrier of nucleic acid, polysaccharide has several advantages over other carrier materials. Since polysaccharide has reactive groups, the chemical modification can be readily made to change the chemical, physical, and biological properties. Another noticeable feature of polysaccharide is to compose of different sugars which can be recognized by the corresponding cell receptors of sugar specificity. This biological recognition not only permits the receptor-specific targeting of agents to the cell, but also accelerates their cell internalization via the receptor-mediated endocytosis. Based on these findings, several researches for the efficient transfection of nucleic acid have been carried out with cationized polysaccharides.

Dimethylaminoethyl dextran (DEAE-dextran) is one of the oldest carriers which have been widely used for the transfection of nucleic acid into mammalian cells cultured[49]. The DEAE-dextran of positive charge can interact with the nucleic acid of negative charge and consequently form the polyion complexes of nucleic acid and DEAE-dextran. It is recognized that the nucleic acid-DEAE dextran complex is adsorbed onto the cell surface through the simple electrostatic interaction force, followed by internalizing into the cells via an endocytosis pathway. The easy procedure is a big advantage of this method, but the cytotoxicity of DEAE-dextran itself and poor reproducibility are practically problematic.

Chitosan itself is positively charged at a high density and interacted with the nucleic acid of negative charge, which is consequently widely used as a carrier of nucleic acid[50,51]. There have been many research results to investigate the effect of physicochemical properties of nucleic acid-chitosan complexes on the efficiency of gene transfection[52-54].

Sakurai and Shinkai discovered that schizophyllan (Figure 1.4J) can form a hetero-triple helix with a single chain of nucleic acid[55] and is used as a carrier of oligonucleotides[56]. An oligoamine-introduced schizophyllan strongly interacted with nucleic acid to improve the stability and schizophyllan covalently linked with octa-arginine (R8) and arginine-glycine-aspartic acid tripeptide (RGD) enhanced the biological effect of antisense oligonucleotide[57].

For polyion complexation with nucleic acid, various cationized dextrans with different molecular weights and cationized extents were prepared by the reductive amination of oxidized dextran. The physicochemical properties of nucleic acid-cationized dextran complexes to enhance the expression level of nucleic acid were systematically optimized[58,59]. A cationized dextran modified with the oleate residue of hydrophobicity enabled nucleic acid to increase the affinity for the cell membrane and consequently enhance the expression of nucleic acid[60].

Cationized polymers as a carrier of nucleic acid modified with a side chain of a polysaccharide have been also reported. Dextran with a molecular weight of 1,500 was modified to polyethyleneimine (PEI). The dextran-grafted PEI with an optimal extent of dextran grafted not only reduced the cytotoxicity of PEI itself, but also enhanced the expression level of nucleic acid to a higher extent than that of original PEI[61]. A poly-L-lysine grafted by hyaluronic acid with a molecular weight of 15,000 enabled a plasmid DNA to target to the sinusoidal endothelial cells expressing a hyaluronic acid receptor[62].

Since stem cells possess their inherent high potentials of proliferation and differentiation into different cell lineages, they have been widely investigated for their clinical applications to induction of tissue regeneration[63]. Since transfection of nucleic acid can manipulate stem cells in terms of the biological functions as well as the proliferation and differentiation abilities, it is necessary to develop the efficient transfection technologies of nucleic acid usable as a tool for the basic research of stem cells biology and medicine. For example, induced pluripotent stem (iPS) cells were prepared by the transfection of nucleic acids for terminally differentiated cells[64]. In addition, stem cells manipulated by nucleic acids to activate and improve the biological functions can be used for cell transplantation therapy. Here, recent research results about the non-viral carrier of cationized polysaccharide for stem cells aiming at the enhancement of induction of tissue regeneration are introduced.

We have explored a cell-specific gene carrier of polysaccharides which can be recognized by the cell surface receptors of sugar-recognition for enhanced expression of nucleic acid. Pullulan (Figure 1.4K), dextran, and mannan (Figure 1.4L) were used as the starting polysaccharides of transfection carrier[65]. To cationize the polysaccharide for formation of polyion complex with plasmid DNA, the spermine of a polyamine present in the body was introduced to the hydroxyl groups of polysaccharide by a carboxydiimidazole activation method. Complexation with the polysaccharide derivatives enabled a plasmid DNA to enhance the expression level of mesenchymal stem cells (MSC) to a significantly high extent compared with that of LipofectAmine 2000® commercially available, while the enhanced gene expression depended on the polysaccharide type (Figure 1.5). The complex of spermine-pullulan and plasmid DNA showed the highest level of gene expression among the plasmid DNA complexes with other cationized polysaccharides. An inhibition assay with asialofetuin which is a ligand of the sugar-recognizable cell surface receptors revealed that the blockage of cell receptors suppressed the level of gene expression. These findings suggest the possibility that the plasmid DNA complex with the cationized polysaccharide derivative is selectively internalized into cells through a sugar-specific receptor of cell surface, resulting in the enhanced gene expression. The similar enhancement of gene expression by the cationized polysaccharide was observed for other cells, such as embryonic stem cells and adipo-derived stromal cells (unpublished data).

MSC are being expected as one of cell sources usable for cardiac reconstruction because of their differentiation potential and ability to supply growth factors. However, the therapeutic potential of MSC is often hindered by the poor viability at the transplanted site. Therefore, as one trial to overcome this issue, a non-viral carrier of cationized polysaccharide is introduced to manipulate MSC for activation of the biological functions[66]. When manipulated by the spermine-dextran complex with plasmid DNA of adrenomedullin (AM), MSC secreted a large amount of AM which is an anti-apoptotic and angiogenic peptide (Figures 1.6A and B). Transplantation of AM-manipulated MSC (AM-MSC) significantly improved cardiac functions after myocardial infarction compared with that of MSC alone (Figure 1.6 C).

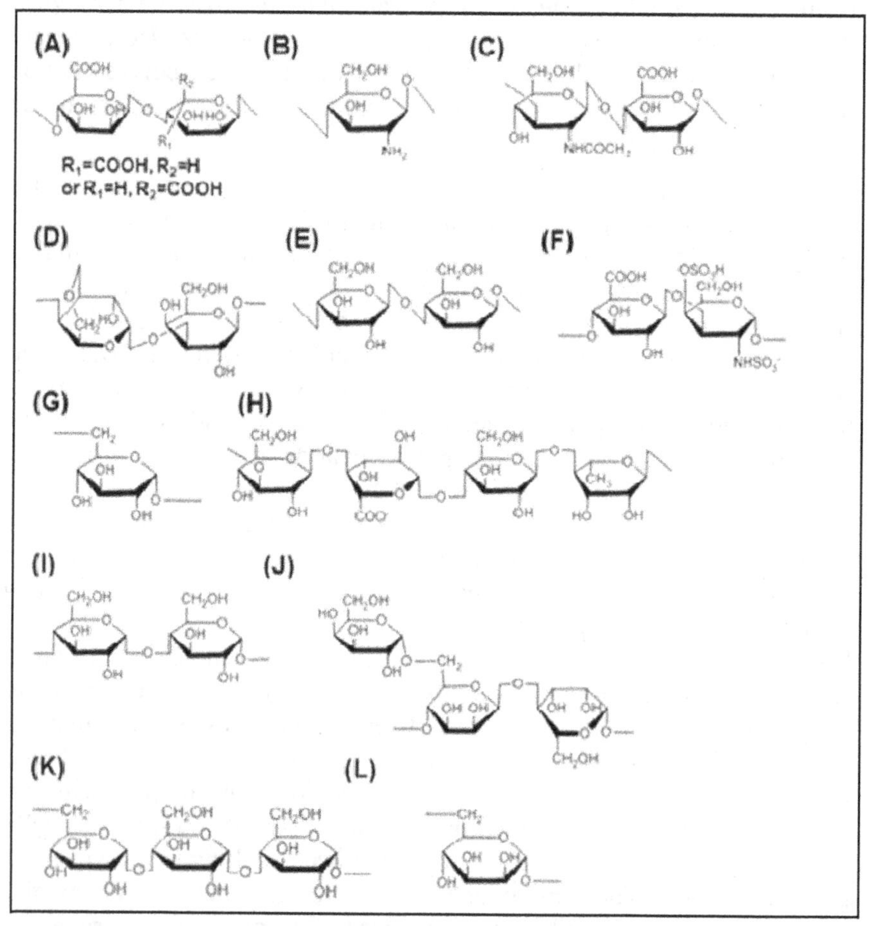

FIGURE 1.4 Chemical structures of polysaccharides used for cell scaffold and drug delivery carrier. (A) alginate, (B) chitosan, (C) hyaluronic acid, (D) agarose, (E) cellulose, (F) chondroitin sulfate, (G) dextran, (H) gellan gum, (I) starch, (J) shizophyllan, (K) pullulan, and (L) mannan.

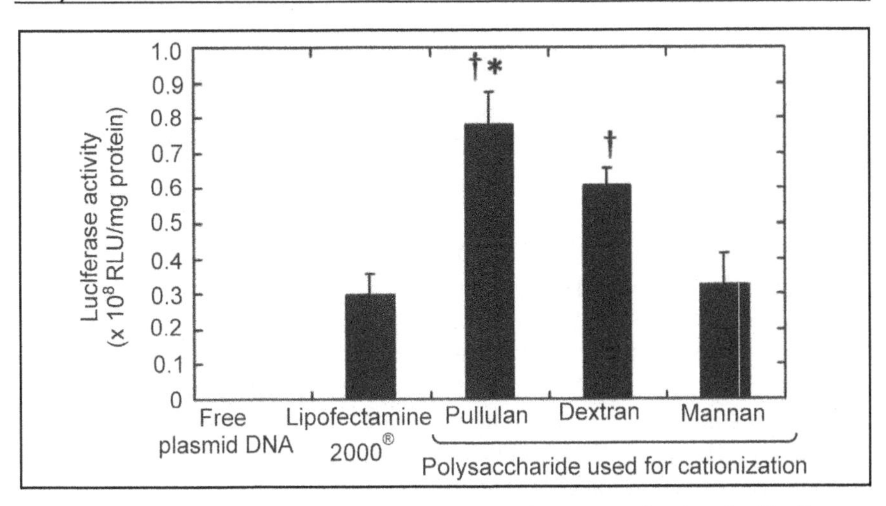

FIGURE 1.5 Effect of polysaccharide types on the expression level of luciferase plasmid DNA complex for MSC. *, p<0.05; versus the expression level of complexes prepared by other spermine-polysaccharides. †, p<0.05; versus the expression level of complexes prepared by Lipofectamine 2000® [65].

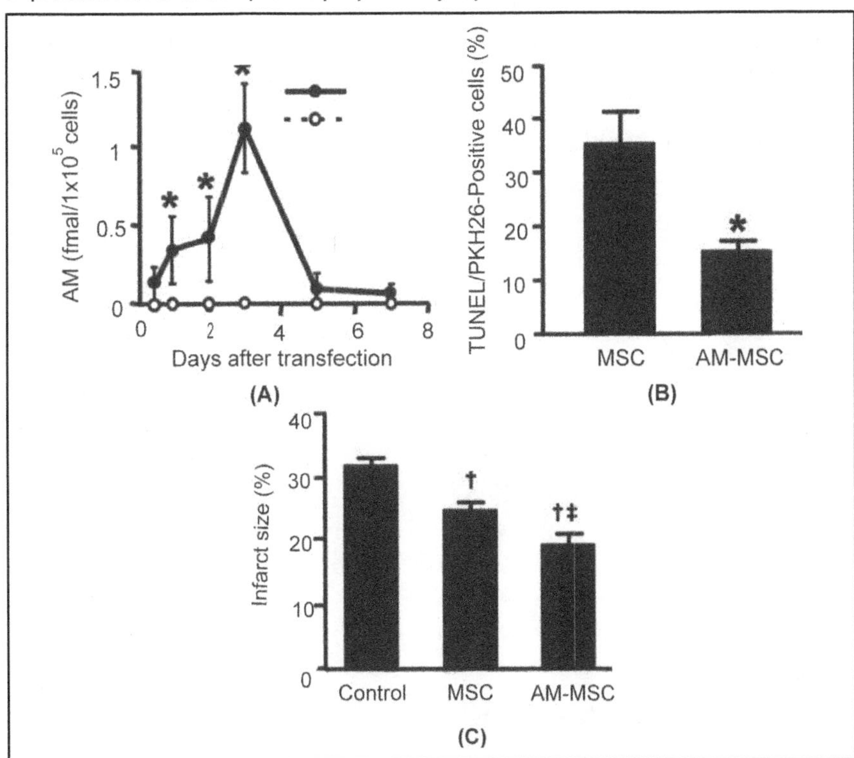

FIGURE 1.6 (A) Time course of AM secreted from MSC following the transfection by spermine-dextran-AM plasmid DNA complexes, AM-MSC. *, $p < 0.05$; versus the

level of original MSC at the corresponding time period. (B) *In vivo* anti-apoptotic effects of AM-MSC. Quantitative analysis of *in vivo* TUNEL assay for AM-MSC and MSC was performed. *, $p < 0.05$; versus the MSC. (C) Therapeutic effects of AM-MSC transplantation on the myocardial infarct size 4 weeks after coronary ligation. †, $p < 0.05$; versus the Control group. ‡, $p < 0.05$; versus the MSC group[66].

RNA interference (RNAi) has been recognized as a phenomenon that messenger RNA (mRNA) is sequence-specifically degraded to suppress the biological function of the corresponding protein[67]. Induction of this RNAi by a siRNA has been scientifically and therapeutically noted in cell biology. The siRNA-based mRNA-specific suppression will be able to artificially enhance or suppress the level of the subsequent gene expression, resulting in the biological manipulation of cell functions. We used spermine-dextran as a transfection carrier of siRNA for the gene suppression of MSC (Figure 1.7A)[68]. It has been reported that MSC are preferably differentiated into osteoblasts rather than adipocytes by a transcription coactivator containing PDZ-binding motif (TAZ) endogenously present[68]. Transfection of TAZ-siRNA complex with spermine-dextran (Figure 1.7B) enabled MSC to promote their differentiation into adipocytes (Figure 1.7C). This is a promising and new technology to control the differentiation direction of cells, which is different from the conventional methodology where the culture medium is modified.

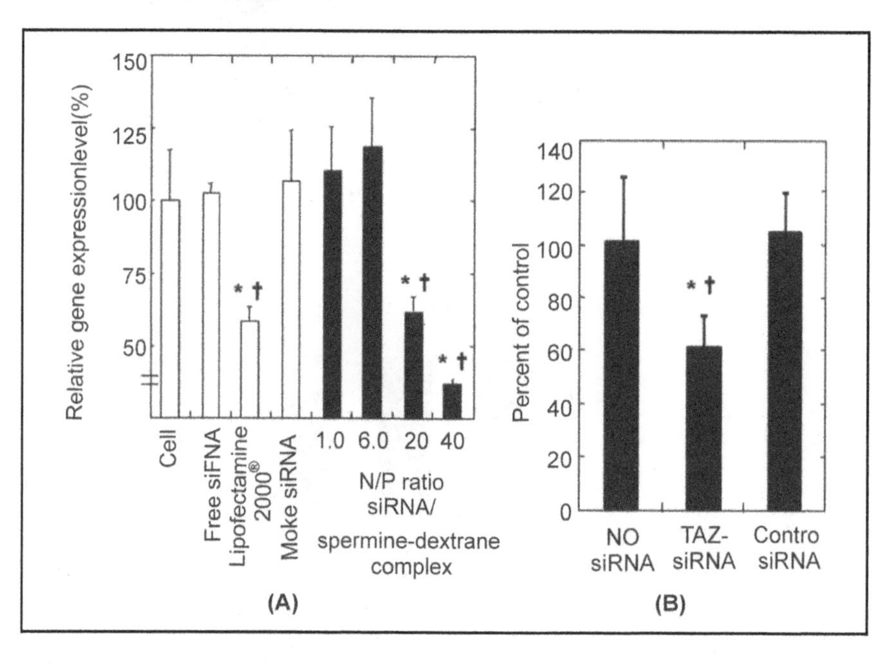

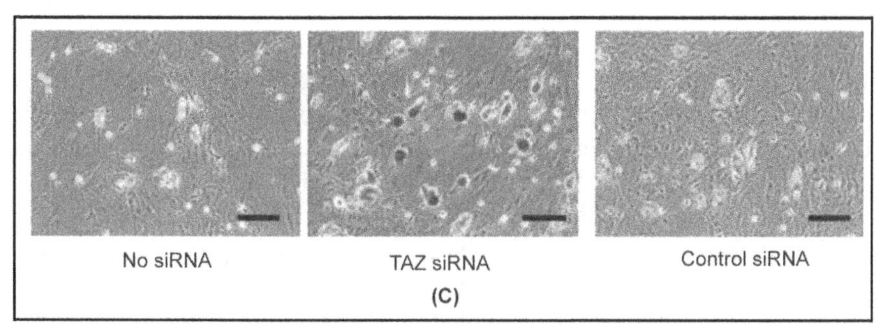

FIGURE 1.7 (A) Suppression effect of spermine-dextran-siRNA complexes on relative gene expression of MSC. As controls, free siRNA, siRNA for non-sense GFP, and Lipofecamine 2000-siRNA complex were used. Percent expressed for MSC without siRNA transfection (Cell) is 100%. *, $p < 0.05$; versus the expression percentage of MSC without siRNA transfection. †, $p < 0.05$; versus the expression percentage of MSC transfected with free siRNA. N/P ratio defines as the molar ratio of nitrogen molecule of spermine-dextran to the phosphorous ones of siRNA. (B) Level change in the TAZ mRNA expression of MSC transfected with spermine-dextran-siRNA complexes. Level of TAZ mRNA expression was evaluated 2 days after siRNA transfection. Amount of siRNA used for transfection is 10 pmole. Percent changed for MSC without siRNA transfection is 100%. *, $p < 0.05$; versus the change percentage of MSC without siRNA transfection. †, $p < 0.05$; versus the change percentage of MSC transfected with the complex of control-siRNA. (C) Phase-contrast microscopic pictures of MSC stained by oil red O. Cells were incubated for 12 days after siRNA transfection. The dose of siRNA used was 20 nM. Bar = 100 μm[68].

When cells are manipulated and used for transplantation therapy, in addition to the enhancement of transfection and expression of nucleic acid, it is undoubtedly important to consider the physiological and functional conditions of cells transfected from the viewpoint of the practical usage. In the conventional procedure of nucleic acid transfection, normally the non-viral carrier is complexed with a plasmid DNA, and then added to the culture medium of cells for transfection. In this case, although the presence of serum is essential to maintain the culture conditions of cells biologically good, the transfection culture is generally being carried out without the serum. This is because the plasmid DNA-carrier complex often interacts with serum components. This interaction often reduces the extent of complex internalized into cells, leading to the suppressed expression level of nucleic acid. Taken together, we cannot always say that the culture condition for the conventional transfection is good in terms of cells viability. Basically, there are two approaches to improve the culture conditions of nucleic acid transfection. One is the technical modification to perform the transfection of nucleic acid even in the presence of serum. The other is to improve the methodology of cell culture which enables cells to physiologically proliferate under good conditions in the transfection culture with nucleic acid. Since a three-dimensional culture substrate has a large surface area available for cell attachment and the subsequent proliferation compared with a two-dimensional tissue culture plate, cells can be generally proliferated in the three-dimensional substrate at higher rates and for longer time periods than those in the two-dimensional one. Moreover, combination with a perfusion culture method can supply nutrients and oxygen to the cells proliferated in the three-dimensional (3D) substrate efficiently compared with a static culture method, while harmful metabolic products and wastes generated from cells can be excluded rapidly. It has been previously demonstrated that the proliferation of MSC was greatly influenced by their culture method and significantly enhanced by the perfusion culture method compared with the static method. Therefore, it is highly expected that combination of cationized polysaccharide-based transfection with the culture method enhances the expression level of nucleic acid to a great extent compared with the conventional method of nucleic acid transfection.

A new non-viral method of transfection was designed to enhance the expression level of nucleic acid for rat MSC. The spermine-pullulan was complexed with a plasmid DNA of luciferase and coated on the surface of culture substrate together with Pronectin® of artificial cell adhesion protein (Figure 1.8A)[70]. MSC were cultured and transfected on the complex-coated substrate (reverse transfection), and the level and duration of gene expression were compared with those of MSC transfected by culturing in the medium containing the plasmid DNA-spermine-pullulan complex (conventional method). The gene expression was enhanced and prolonged by the reverse transfection method to a significantly great extent compared with that of the conventional method (Figure 1.8B). The reverse method permitted the transfection culture of MSC in the presence of serum, in marked contrast to the conventional method, which gave cells a good culture condition, resulting in lower cytotoxicity (Figure 1.8C).

The reverse transfection can be carried out for 3D culture substrate from the viewpoint of transfection procedure. Non-woven fabic of polyethylene terephtalate (PET) coated with the complex and Pronectin® while the cell culture was performed by bioreactor systems, such as an agitated and stirring culture methods. The level and duration of gene expression for MSC were significantly enhanced by the two bioreactor methods compared with that of the static method (Figure 1.8D). It is possible that the medium circulation improves the culture conditions of cells in terms of oxygen and nutrition supply and wastes excretion, resulting in an enhanced nucleic acid expression. These results strongly demonstrated that comparable to the research and development of non-viral carriers themselves, it is important to improve the technology and methodology of cell culture which give cells good conditions to maintain their vital and biological functions as well as efficiently enhance nucleic acid transfection[71].

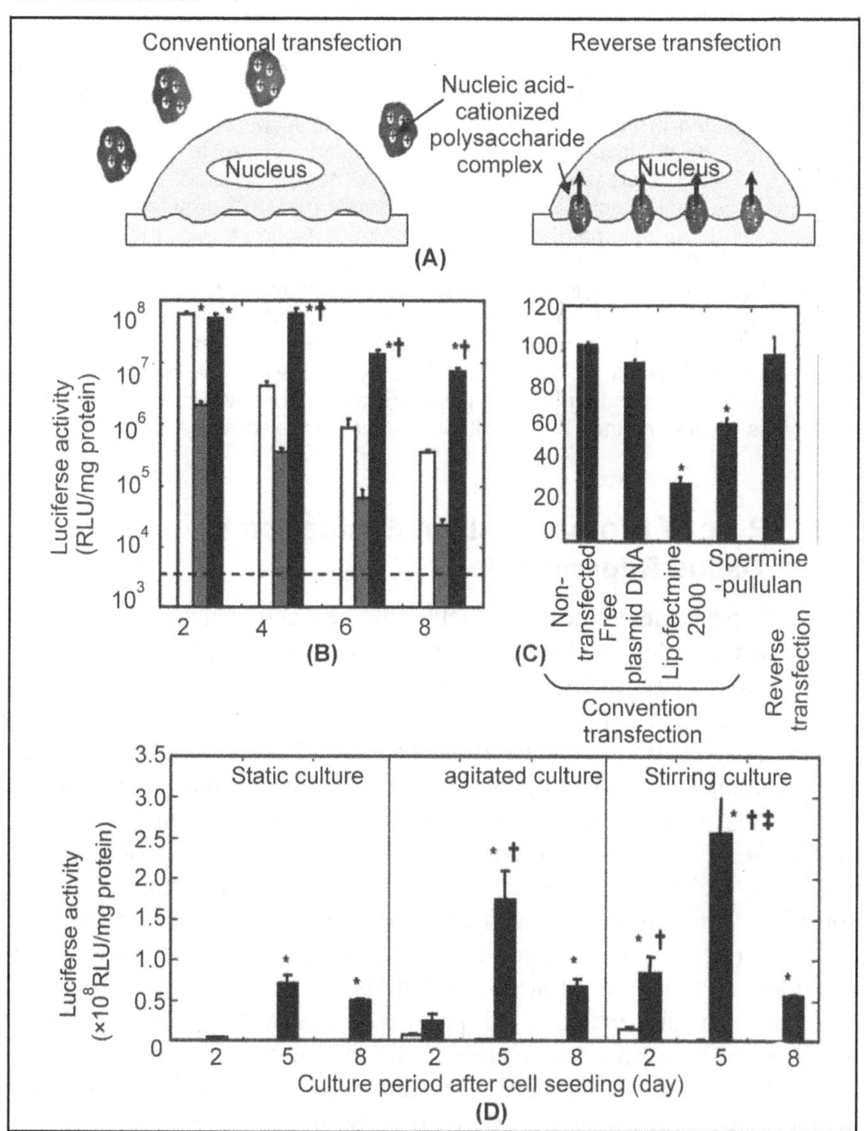

FIGURE 1.8 (A) Illustration of conventional or reverse gene transfection. (B) Time course of luciferase expression level of MSC transfected by the

conventional (open and light gray columns) and reverse methods (solid columns) in the static cuture: (open columns) the plasmid DNA-spermine-pullulan complex in the absence of fetal calf serum (FCS), (light gray columns) the plasmid DNA-spermine-pullulan complex in the presence of FCS, and (solid columns) the plasmid DNA-spermine-pullulan complex in the presence of FCS. The dotted line indicates the level of non-transfected, original MSC. *, $p < 0.05$; versus the level in the presence of FCS by the conventional method at the corresponding time. †, $p < 0.05$; versus the level in the absence of FCS by the conventional method at the corresponding time. (C) Cell viability of MSC 2 days after conventional and reverse transfection cultures. The cells were transfected by the conventional method with free plasmid DNA or that complexed with Lipofectamine 2000 and spermine-pullulan in the absence of serum. The cells were transfected by the reverse method with plasmid DNA-spermine-pullulan complex in the presence of serum. *, $p < 0.05$; versus the cell viability of non-transfected, original MSC. (D) Time course of luciferase expression level of MSC transfected by the conventional (open columns) and reverse methods (solid columns) in the static, agitation, and stirring cultures in the PET non-woven fabric: (open columns) the plasmid DNA-spermine-pullulan complex in the absence of FCS and (solid columns) the complex in the presence of FCS. *, $p < 0.05$; versus the level in the presence of FCS by the conventional method at the corresponding time. †, $p < 0.05$; versus the level in the absence of FCS by the reverse method in the static culture at the corresponding time. ‡, $p < 0.05$; versus the level in the absence of FCS by the reverse method in the agitation culture at the corresponding time[70].

1.3 Role of Biomaterials in Evaluation of Tissue Regeneration

It is indispensable to develop methodologies and technologies which enable to accurately and non-invasively evaluate the phenomenon and process of tissue regeneration. The distribution and biological function of cells and the ECM in the site to be regenerated should be revealed during the process of tissue regeneration process. There is no doubt that methodologies and technologies of molecular imaging play an important role in the evaluation of tissue regeneration. Molecular imaging is defined as the *in vivo* visualization of spatial and temporal distribution of molecular biological processes in the cell and tissue of interest[72]. The non-invasiveness of molecular imaging enables to repeatedly investigate the status of tissue regeneration induction in a patient, which gives more effective and active strategy individually from results obtained. Molecular imaging includes the imaging probes and the corresponding imaging modality. So far, various types of imaging probe have been reported, such as Gd^{3+}, Mn^{2+}, ^{19}F, and iron oxide nanoparticles for magnetic resonance imaging (MRI), radioisotopes (^{99m}Tc, ^{111}In, ^{123}I, ^{18}F, ^{64}Cu, and ^{124}I) or their derivatives for positron emission tomography (PET) and single photon emission computed tomography (SPECT), luciferase and β-galactosidase for bio-luminescence imaging, and quantum dots and near-infrared fluorescent (NIRF) dyes for fluorescence imaging. There are two technologies required for the evaluation of tissue regeneration (Figure 1.9);

1. To sequentially trace the *in vivo* fate and biological function of cells transplanted and
2. To accurately visualize the phenomenon and process of tissue regeneration by using a specific molecule expressed or secreted during the regeneration process.

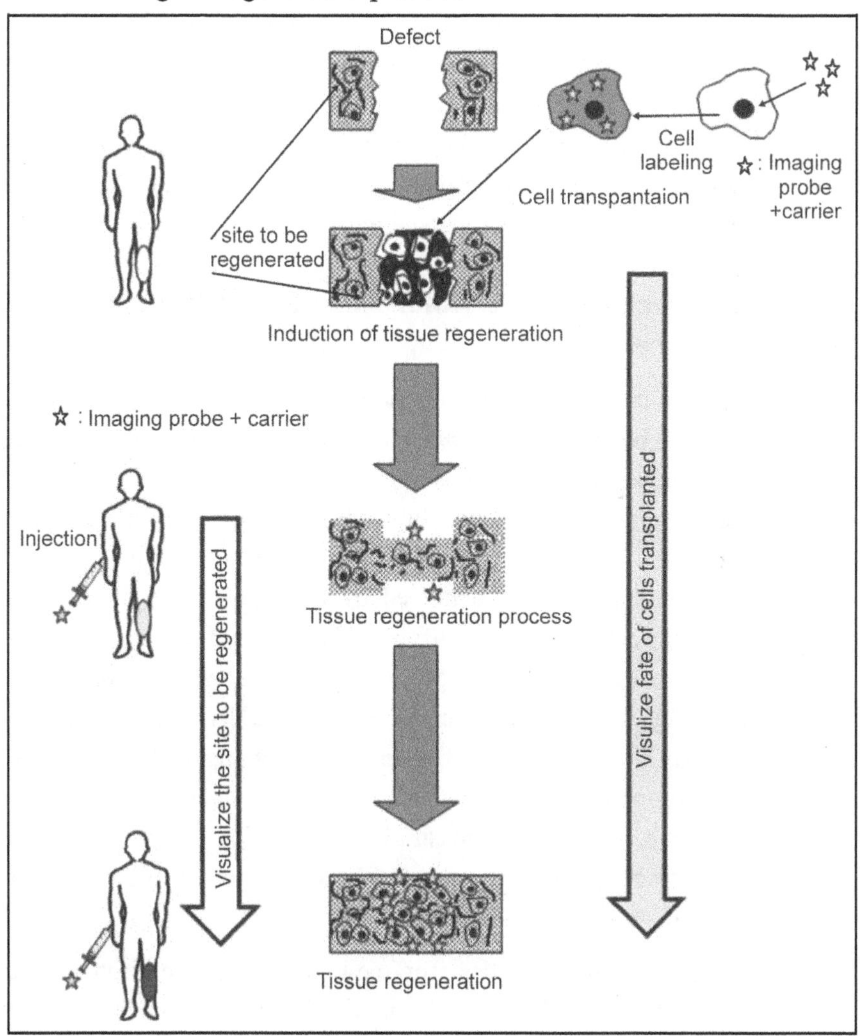

FIGURE 1.9 Technologies required for evaluation of tissue regeneration.

These can be achieved by the internalization of imaging probes into the cells (cell labeling) or their delivery to the site regenerated. Therefore, the DDS technologies of accelerated permeation and absorption or targeting of drug can be utilized to maximize the efficacy of molecular imaging by increasing in the ratio of signal to noise (S/N). It has been demonstrated that combination with nano-sized carriers, such as water-soluble polymers, polymer micelles, emulsions, and liposomes, allows the imaging probes to simultaneously condense and to stably deliver in the blood circulation, resulting in an enhanced imaging efficacy[73,74]. Furthermore, modification with targeting moieties such as antibody or targeting peptide enables nano-sized carriers to be actively delivered to the tissue of interest. On the basis of these findings, the nano-sized carriers incorporating imaging probes are highly expected to play an important role in the evaluation of tissue regeneration.

1.3.1 Polysaccharide as a Nano-sized Carrier for Cell Labeling

There have been many research trials on cell labeling technologies with imaging probe. The imaging probes used for cell labeling include iron oxide nanoparticles, several radioisotope derivatives, luciferase and β-galactosidase, quantum dots, and fluorescent dye (green fluorescence protein (GFP) etc.). In the case of the iron oxide nanoparticles or quantum dots, cell labeling is achieved by making use of transfection reagent, external stimulation or modifying the surface of particles suitable for cellular internalization. In the case of luciferase, β-galactosidase, and GFP, tranfection with the expressing plasmid DNA enable cells to visualize by bio-luminescence or fluorescence imaging methods. In the case of radioisotope derivatives, cell labeling is achieved by in advance transfection for cells with the plasmid DNA coding a protein which specifically interacts with a derivative and the subsequent incubation of the cells with the derivative.

Among the imaging probes for cell labeling, the iron oxide nanoparticles have been the most frequently used[75-77]. The iron oxide nanoparticles have been paid great attention as a material for biomedical applications[78], such as a carrier for magnet responsive drug delivery system, a heating substance for hyperthermia, an adsorbent for magnetic separation and selection columns, and a negative contrast agent for MRI. MRI has been used for clinical diagnosis, while it is expected to be a molecular imaging modality with advantages of high spatial resolution and penetration depth. It is practically necessary for efficient cell labeling to accurately control the size and surface state of iron oxide nanoparticles. It has been well recognized that the iron oxide nanoparticles without any surface coatings often suffer from their aggregation in water or tissue fluid[79], resulting in the low labeling efficiency. Polymer coating enables iron oxide nanoparticles to make stable dispersion, and the iron oxide nanoparticles coated have been used in clinic. Since the iron oxide nanoparticles cannot be internalized into cells because of their surface nature of neutral or negative charge. Under these conditions, the cell labeling is performed by the combination of gene transfection agents or physical stimuli described above. However, such a labeling procedure is problematic in terms of its complexity and low efficiency. Therefore, a novel labeling procedure with simple and short-term incubation is highly expected. On the other hand, there are three important points for *in vivo* tracing of cell labeled; low cytotoxicity, maintenance of cells function, and long retention in the cells.

We designed iron oxide nanoparticles with different sizes and surface potentials by the conventional co-precipitation of ferric and ferrous ions in the presence of pullulan derivatives[80]. The size and surface potential of iron oxide-pullulan nanoparticles were changed by altering the mixing molar ratio of pullulan OH groups to ferric ions and the mixing percentage of pullulan derivatives, respectively. When MSC were labeled with the iron oxide-pullulan nanoparticles by the co-culture for 1 hr, the labeling efficiency and ^{1}H-MRI relaxivity were greatly influenced by the particle size and surface potential (Figures 1.10A and B), while the labeling procedure did not affect the viability (Figure 1.10C) and differentiation (Figure 1.10D) ability of MSC. These findings indicate that iron oxide-pullulan nanoparticle is a promising tool for the MRI labeling of MSC.

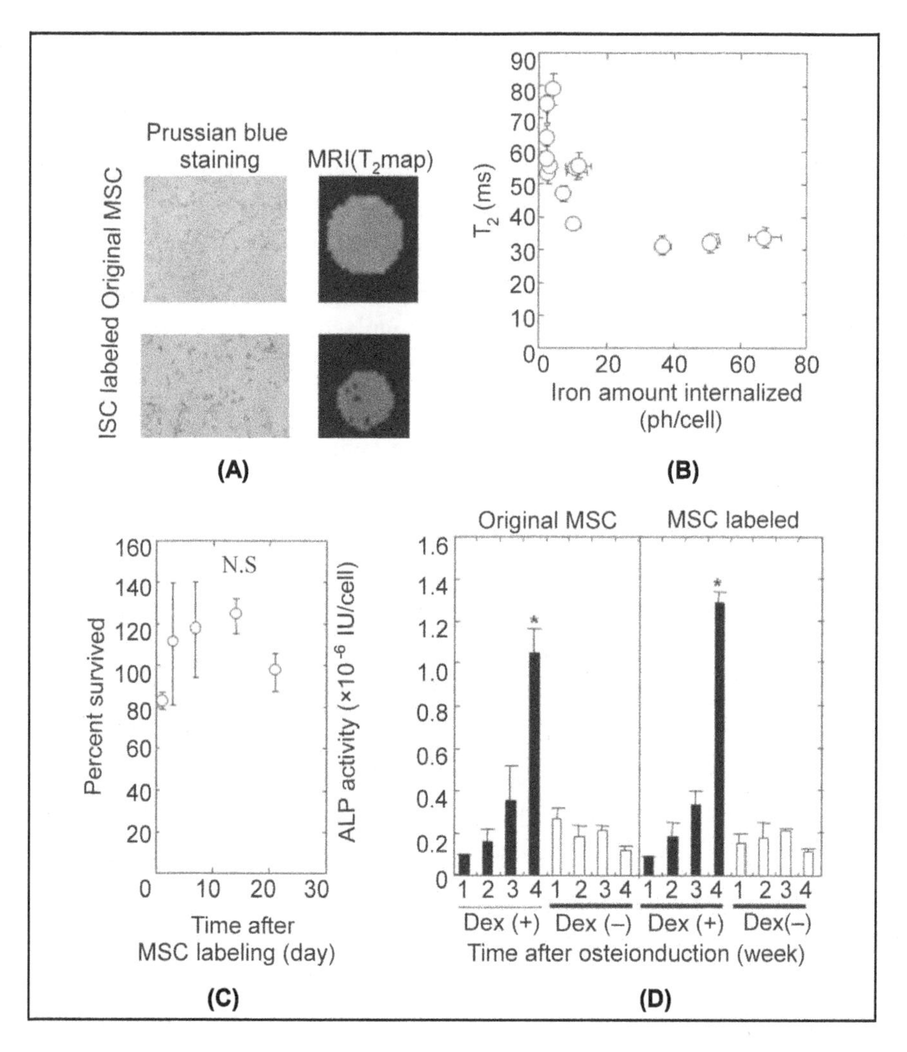

FIGURE 1.10 (A) Representative photographs of Prussian blue staining or quantitative T_2-map (MRI) for MSC labeled with or without iron oxide-pullulan nanoparticles. MSC are embedded in 10 wt% gelatin (200 μl) at a density of 1×10^6 cells/ml for MRI acquisition. (B) Relationship between the iron amount internalized and the T_2 relaxation time. (C) Time profile of viability of MSC labeled with iron oxide-pullulan nanoparticles. N.S.; not significant among times after MSC labeling. (D) Time profile of alkaline phosphatase (ALP) activity of original MSC or MSC labeled with iron oxide-pullulan nanoparticle in the medium containing 28 mM L-ascorbic acid, 10 mM β-glycerophosphate with (Dex (+)) or without 10 nM dexamethasone (Dex (-)). *, $p < 0.05$; versus the original MSC or MSC labeled with iron oxide-pullulan nanoparticles without induction of osteogenic differentiation at the corresponding time[80].

1.3.2 Polysaccharide as a Nano-sized Carrier for Targeting to the Regeneration Site

As described above, it has been demonstrated that nano-sized carriers are a powerful tool to effectively deliver imaging probes to the tissue of interest. There are a lot of research results about the visualization of tumor sites[81,82] or arteriosclerosis[83,84] by the injection of nano-sized carriers incorporating the related imaging probe. However, there are only a few researches on the visualization of regeneration site by making use of imaging probes combined with nano-sized carriers[85-88].

Polysaccharides and their derivatives have been paid great attention for drug delivery approaches because of their availability to improve the pharmacokinetical and pharmacodynamical characteristics of therapeutic agents, such as low molecular weight drugs, bioactive peptides or proteins, enzymes, and plasmid DNA. The chemical or physical conjugation of therapeutic agents with the polysaccharide derivatives increases their biological stability in the blood circulation, and consequently prolong the time period of activity. On the other hand, polysaccharides have been used to target the agents to the tumor (passive targeting) or the liver (active targeting). In addition, paramagnetic substance-loaded or radiolabeled polysacchrides have been applied for diagnostic imaging. Therefore, it is highly expected that imaging probes conjugated with the polysaccharide derivatives enhance the visualization efficacy in the site to be regenerated.

Angiogenesis is defined as the generation of capillaries from natural vessels and is a fundamental process involved in various phenomena including development, wound healing, tissue regeneration (physiological angiogenesis), and the progression of chronic inflammation and tumor (pathological angiogenesis)[89]. Angiogenesis occurs via several processes:

(i) the degradation of extracellular matrix surrounding the existing vasculature;
(ii) the proliferation and migration of endothelial cells thereat as well as the attraction of blood-derived macrophages and circulating stem cells; and
(iii) the integration of endothelial cells, followed by tube formation[90)].

The hypoxia-inducible factor (HIF)-1α, vascular endothelial growth factor (VEGF), matrix metalloproteinase (MMP), and $\alpha_v\beta_3$ integrin have been used as a molecule of interest. Currently, the clinical trial on the induction therapy of angiogenesis is being performed for patients with ischemic diseases by cell transplantation or administration of bioactive substances based on the controlled release system of gelatin hydrogels[91]. However, often the angiogenesis is not detected angiographically although the disease appearance of patients got better. This is mainly because there are no diagnostic systems to clinically detect blood vessels with a small diameter regenerated.

A new polysaccharide-based imaging probe have been designed and prepared for the MRI evaluation of therapeutic angiogenesis[92]. The probe consists of dextran, diethylenetriaminepentaacetic acid (DTPA) residue of a chelator, Gd^{3+}, and cyclic peptide containing an arginine-glycine-aspartic acid sequence (cRGD) with an inherent affinity for the $\alpha_v\beta_3$ integrin (cRGD-dextran-DTPA-Gd, Figure 1.11A). The cRGD-dextran-DTPA-Gd had an affinity for cells expressing the $\alpha_v\beta_3$ integrin and showed a higher longitudinal relaxivity compared with the DTPA-Gd of MRI contrast agent clinically used. Right femoral, external iliac, and deep femoral and circumflex arteries and veins were surgically ligated to prepare a mouse model of hindlimb ischemia. A laser Doppler analysis and histological evaluation experimentally confirmed that the hindlimb ischemia was naturally healed accompanying angiogenesis, while the $\alpha_v\beta_3$ integrin was expressed in the ischemic-angiogenic region without any treatment. When intravenously injected into mice with hindlimb ischemia, the cRGD-dextran-DTPA-Gd were significantly accumulated in the ischemic-angiogenic region and showed an MR ability to detect the ischemic-angiogenic region (Figures 1.11B-D).

Non-invasive imaging technologies have been widely used in clinical diagnosis. However, each imaging modality is based on quite different principles, and has the advantages and disadvantages. Generally, a single technique does not always correspond to all the requirements for diagnosis imaging[93]. As one trial to tackle the issue, it is practically possible to design a multimodal imaging system. A combinational imaging system composed of different imaging modalities may compensate the deficiencies of single imaging modality, while it gives useful and new tools to biomedical researches and clinical diagnosis. Currently, some prototypes of multimodal imaging system including MRI–optical, NIRF–SPECT, PET-CT, and SPECT-MRI, have been introduced. However, the research and development of multimodal imaging probes are still in an early stage although they are highly required to realize the idea of multimodal imaging. To design and prepare such a multimodal imaging system, it is also of prime necessity to make the best of DDS technology to enhance the accumulation of imaging agents in the target tissue, resulting in the increased S/N ratio.

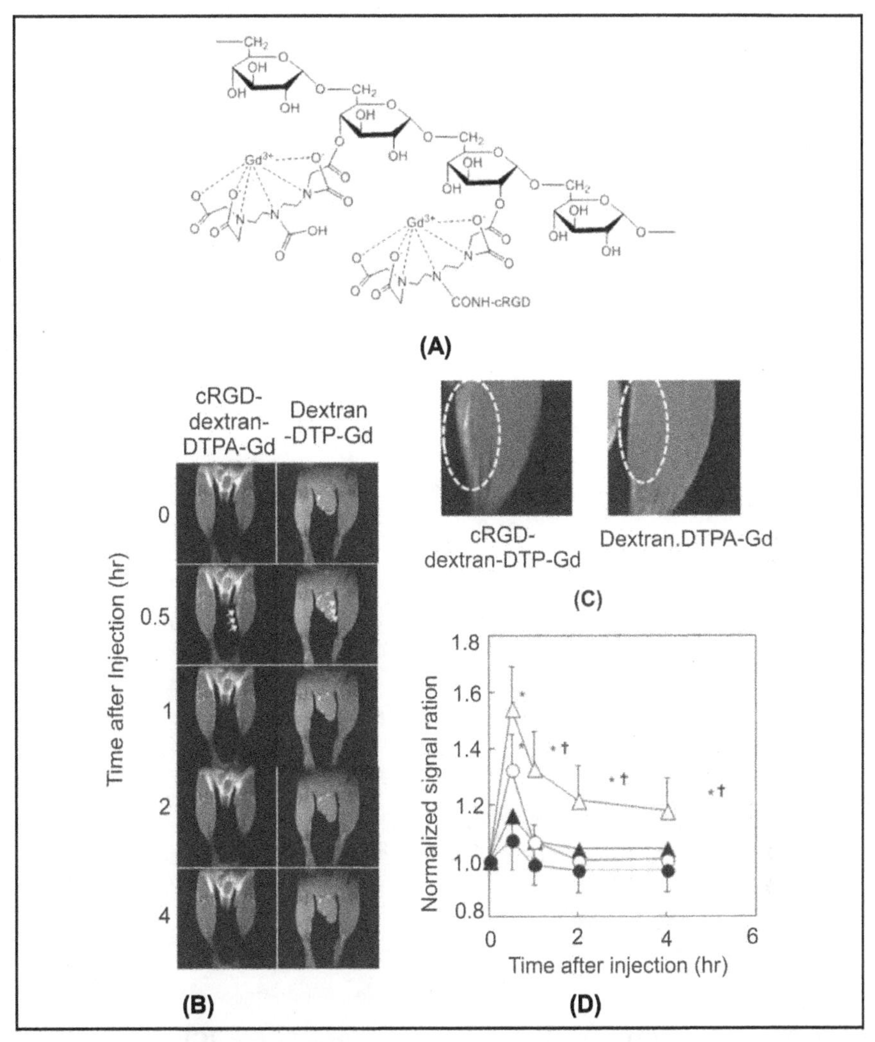

FIGURE 1.11 (A) Chemical structure of cRGD-dextran-DTPA-Gd. (B) Representative MRI pictures of mice hindlimb region before and after injection of dextran-DTPA-Gd or cRGD-dextran-DTPA-Gd. The right side shows the ischemic hindlimb. Arrows indicate the ischemic-angiogenic region. (C) Representative enlarged image of ischemic-angiogenic region (dashed ellipses). (D) Time profiles of T_1 signal intensity ratio in the mice hindlimb after injection of dextran-DTPA-Gd (○, •) or cRGD-dextran-DTPA-Gd (Δ, ▲). Signals (5 points each) were acquired in normal (solid) and ischemic-angiogenic region (open). *, $p < 0.05$; versus the normalized signal ratio in the normal region after the injection of corresponding agent at the corresponding time. †, $p < 0.05$; versus the normalized signal ratio in the ischemic-angiogenic region after injection of dextran-DTPA-Gd at the corresponding time[92].

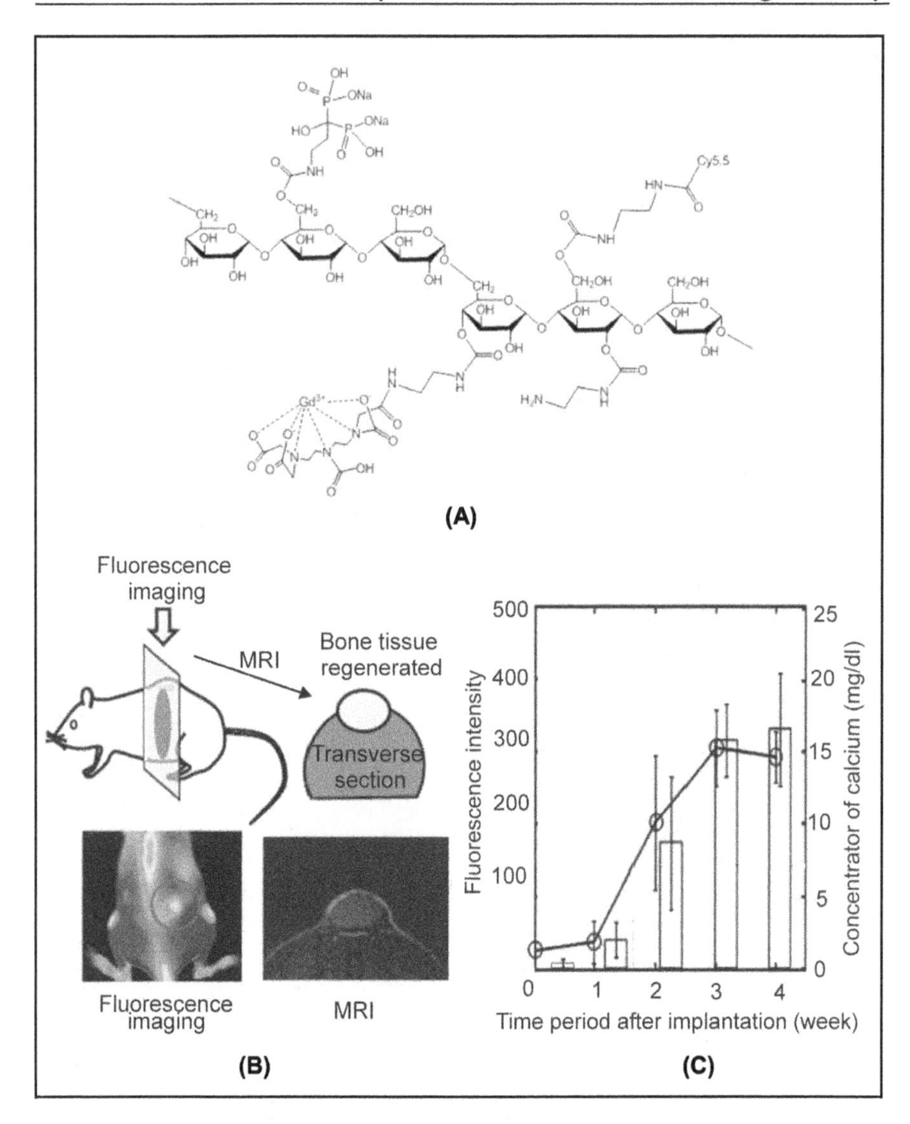

FIGURE 1.12 (A) Chemical structure of PA-pullulan-F/M. (B) Representative pictures of mice injected with PA-pullulan-F/M acquired by fluorescence imaging or MRI. Injection was performed for mice 3 weeks after the hydrogel implantation. Site implanted with hydrogel incorporating BMP-2 is indicated by a red circle for the fluorescence image. Trans-axial multi-slice T_1-weighted MR images were repeatedly acquired with a spin-echo sequence. (C) Time profiles of calcium concentration of subcutaneous tissues around gelatin hydrogels incorporating BMP-2 implanted (open bars) and fluorescent intensity of the site around gelatin hydrogels incorporating BMP-2 implanted (open circles)[96].

Generally, bone healing and repairing have been diagnosed so far by CT, X-ray, and MRI, and their methodology has obtained the clinical reliability. However, the convention imaging methods cannot always give us detailed information about soft and bone tissues regenerated even in the early stage. Although several researches have been performed on bone-specific imaging probes[94,95], few multimodal imaging probes are developed. If a multimodal imaging system to visualize the regeneration process of bone tissues can be developed, the extent of bone regeneration and repairing will be more clearly diagnosed and consequently become a new therapeutic strategy with high reliability. In addition, the system can also give an evaluation method to observe whether or not the process of bone tissue regeneration takes place properly. For this purpose, the modality combination of MRI with a superior property of tomography as well as a high spatial resolution and optical imaging with a high sensitivity and specificity would be a promising choice.

A new polysaccharide-based imaging probe have been designed and prepared for multimodal imaging system for the evaluation of bone regeneration processes[96]. The polysaccharide-based imaging probe consists of pullulan, DTPA, Gd^{3+}, Cy5 of fluorescent dye, and pamidronate (PA) of bisphosphonates with a high affinity for hydroxyapatite (PA-pullulan-F/M, Figure 1.12A). The PA-pullulan-F/M had an affinity for hydroxyapatite and showed an MRI ability similar to Gd-DTPA clinically used. A gelatin hydrogel incorporating bone morphogentic protein (BMP)-2 was prepared and implanted subcutaneously into mice to obtain an animal model of bone regeneration[97]. When intravenously injected into mice with the bone tissue ectopically formed by the BMP-2-incorporated hydrogel to evaluate their body distribution by the fluorescence imaging and MRI, the PA-pullulan-F/M accumulated in the bone tissue regenerated (Figure 1.12B). It should be noted that the time profile of fluorescent intensity well corresponded with that of calcium amount in the bone tissue newly formed (Figure 1.12C). These findings clearly indicated that the PA-pullulan-F/M is a useful multimodal imaging probe which enables to evaluate not only the phenomenon, but also the process of bone regeneration.

Conclusion

This chapter described the role of polysaccharides in the induction and evaluation of tissue regeneration and summarized the present status of related researches. The advantages of polysaccharide as a material include the ease of acquisition, chemical modification, and construct formation. With the development of "Glycoscience", the novel biological functions of polysaccharides are being revealed one after another. It is no doubt that innovative polysaccharide constructs for biomedical applications containing tissue regeneration technologies will be created by actively making use of these functions. The tissue regeneration is interdisciplinary research field. To achieve effective induction and evaluation using polysaccharide-based technologies, substantial collaborative researches between material, pharmaceutical, image technological, biological, and clinical scientists are highly required. Such researchers must have knowledge in medicine, biology, pharmacology, image technology in addition to material sciences. We will be happy if this chapter stimulates readers' interest in the research field of tissue regeneration and assists their understanding of importance in induction and evaluation of tissue regeneration with polysaccharide-based cell scaffold and DDS.

References

1. Langer, R.; Vacanti, J.P. Tissue engineering. Science 1993, 260, 920-6.
2. Mano, J.F.; Silva, G.A.; Azevedo, H.S.; Malafaya, P.B.; Sousa, R.A.; Silva, S.S.; Boesel, L.F.; Oliveira, J.M.; Santos, T.C.; Marques, A.P.; Neves, N.M.; Reis, R.L. Natural origin biodegradable systems in tissue engineering and regenerative medicine: present status and some moving trends. J R Soc Interface 2007, 4, 999-1030.
3. Liu, C.; Zhang, N. Nanoparticles in gene therapy principles, prospects, and challenges. Prog Mol Biol Transl Sci 2011, 104, 509-62.
4. Itaka, K.; Kataoka, K. Progress and prospects of polyplex nanomicelles for plasmid DNA delivery. Curr Gene Ther 2011, 11, 457-65.
5. Singha, K.; Namgung, R.; Kim, W.J. Polymers in small-interfering RNA delivery. Nucleic Acid Ther 2011, 21, 133-47.
6. Xiong, F.; Mi, Z.; Gu, N. Cationic liposomes as gene delivery system: transfection efficiency and new application. Pharmazie 2011, 66, 158-64.
7. Tros de Ilarduya, C.; Sun, Y.; Duzgunes, N. Gene delivery by lipoplexes and polyplexes. Eur J Pharm Sci 2010, 40, 159-70.
8. Ewert, K.K.; Zidovska, A.; Ahmad, A.; Bouxsein, N.F.; Evans, H.M.; McAllister, C.S.; Samuel, C.E.; Safinya, C.R. Cationic liposome-nucleic acid complexes for gene delivery and silencing: pathways and mechanisms for plasmid DNA and siRNA. Top Curr Chem 2010, 296, 191-226.

9. Gaidamakova, E.K.; Backer, M.V.; Backer, J.M. Molecular vehicle for target-mediated delivery of therapeutics and diagnostics. J Control Release 2001, 74, 341-7.
10. Niidome, T.; Huang, L. Gene therapy progress and prospects: nonviral vectors. Gene Ther 2002, 9, 1647-52.
11. Tachibana, R.; Harashima, H.; Shinohara, Y.; Kiwada, H. Quantitative studies on the nuclear transport of plasmid DNA and gene expression employing nonviral vectors. Adv Drug Deliv Rev 2001, 52, 219-26.
12. Lippiello, L. Glucosamine and chondroitin sulfate: biological response modifiers of chondrocytes under simulated conditions of joint stress. Osteoarthritis Cartilage 2003, 11, 335-42.
13. Balakrishnan, B.; Jayakrishnan, A. Self-cross-linking biopolymers as injectable *in situ* forming biodegradable scaffolds. Biomaterials 2005, 26, 3941-51.
14. Liao, I.C.; Wan, A.C.; Yim, E.K.; Leong, K.W. Controlled release from fibers of polyelectrolyte complexes. J Control Release 2005, 104, 347-58.
15. Pongjanyakul, T.; Puttipipatkhachorn, S. Xanthan-alginate composite gel beads: molecular interaction and *in vitro* characterization. Int J Pharm 2007, 331, 61-71.
16. Gigante, A.; Bevilacqua, C.; Cappella, M.; Manzotti, S.;Greco, F. Engineered articular cartilage: influence of the scaffold on cell phenotype and proliferation. J Mater Sci Mater Med 2003, 14, 713-6.
17. Gaissmaier, C.; Fritz, J.; Krackhardt, T.; Flesch, I.; Aicher, W.K.; Ashammakhi, N. Effect of human platelet supernatant on proliferation and matrix synthesis of human articular chondrocytes in monolayer and three-dimensional alginate cultures. Biomaterials 2005, 26, 1953-60.
18. Awad, H.A.; Wickham, M.Q.; Leddy, H.A.; Gimble, J.M.; Guilak, F. Chondrogenic differentiation of adipose-derived adult stem cells in agarose, alginate, and gelatin scaffolds. Biomaterials 2004, 25, 3211-22.
19. Heywood, H.K.; Bader, D.L.; Lee, D.A. Glucose concentration and medium volume influence cell viability and glycosaminoglycan synthesis in chondrocyte-seeded alginate constructs. Tissue Eng 2006, 12, 3487-96.
20. Dausse, Y.; Grossin, L.; Miralles, G.; Pelletier, S.; Mainard, D.; Hubert, P.; Baptiste, D.; Gillet, P.; Dellacherie, E.; Netter, P.; Payan, E. Cartilage repair using new polysaccharidic biomaterials: macroscopic, histological and biochemical approaches in a rat model of cartilage defect. Osteoarthritis Cartilage 2003, 11, 16-28.
21. Tilakaratne, H.K.; Hunter, S.K.; Andracki, M.E.; Benda, J.A.; Rodgers, V.G. Characterizing short-term release and neovascularization potential of multi-protein growth supplement delivered via alginate hollow fiber devices. Biomaterials 2007, 28, 89-98.

22. Vogelin, E.; Baker, J.M.; Gates, J.; Dixit, V.; Constantinescu, M.A.; Jones, N.F. Effects of local continuous release of brain derived neurotrophic factor (BDNF) on peripheral nerve regeneration in a rat model. Exp Neurol 2006, 199, 348-53.
23. Gruber, H.E.; Fisher, E.C., Jr.; Desai, B.; Stasky, A.A.; Hoelscher, G.; Hanley, E.N., Jr. Human intervertebral disc cells from the annulus: three-dimensional culture in agarose or alginate and responsiveness to TGF-beta1. Exp Cell Res 1997, 235, 13-21.
24. Simpson, N.E.; Stabler, C.L.; Simpson, C.P.; Sambanis, A.; Constantinidis, I. The role of the CaCl2-guluronic acid interaction on alginate encapsulated beta TC3 cells. Biomaterials 2004, 25, 2603-10.
25. Yan, L.P.; Wang, Y.J.; Ren, L.; Wu, G.; Caridade, S.G.; Fan, J.B.; Wang, L.Y.; Ji, P.H.; Oliveira, J.M.; Oliveira, J.T.; Mano, J.F.;Reis, R.L. Genipin-cross-linked collagen/chitosan biomimetic scaffolds for articular cartilage tissue engineering applications. J Biomed Mater Res A 2010, 95, 465-75.
26. Okamoto, Y.; Yano, R.; Miyatake, K.; Tomohiro, I.; Shigemasa, Y.; Minami, S. Effects of chitin and chitosan on blood coagulation. Carbohyd Polym 2003, 53, 337-42.
27. Jayakumar, R.; Prabaharan, M.; Sudheesh Kumar, P.T.; Nair, S.V.; Tamura, H. Biomaterials based on chitin and chitosan in wound dressing applications. Biotechnol Adv 2011, 29, 322-37.
28. Miranda, S.C.; Silva, G.A.; Mendes, R.M.; Abreu, F.A.; Caliari, M.V.; Alves, J.B.; Goes, A.M. Mesenchymal stem cells associated with porous chitosan-gelatin scaffold: a potential strategy for alveolar bone regeneration. J Biomed Mater Res A 2012, 100, 2775-86.
29. Shi, H.; Han, C.; Mao, Z.; Ma, L.; Gao, C. Enhanced angiogenesis in porous collagen-chitosan scaffolds loaded with angiogenin. Tissue Eng Part A 2008, 14, 1775-85.
30. Yi, X.; Jin, G.; Tian, M.; Mao, W.; Qin, J. Porous chitosan scaffold and ngf promote neuronal differentiation of neural stem cells *in vitro*. Neuro Endocrinol Lett 2011, 32, 705-10.
31. Soranzo, C.; Renier, D.; Pavesio, A. Synthesis and characterization of hyaluronan-based polymers for tissue engineering. Methods Mol Biol 2004, 238, 25-40.
32. Aigner, J.; Tegeler, J.; Hutzler, P.; Campoccia, D.; Pavesio, A.; Hammer, C.; Kastenbauer, E.; Naumann, A. Cartilage tissue engineering with novel nonwoven structured biomaterial based on hyaluronic acid benzyl ester. J Biomed Mater Res 1998, 42, 172-81.
33. Turner, N.J.; Kielty, C.M.; Walker, M.G.; Canfield, A.E. A novel hyaluronan-based biomaterial (Hyaff-11) as a scaffold for endothelial cells in tissue engineered vascular grafts. Biomaterials 2004, 25, 5955-64.

34. Hemmrich, K.; von Heimburg, D.; Rendchen, R.; Di Bartolo, C.; Milella, E.; Pallua, N. Implantation of preadipocyte-loaded hyaluronic acid-based scaffolds into nude mice to evaluate potential for soft tissue engineering. Biomaterials 2005, 26, 7025-37.
35. Gupta, D.; Tator, C.H.; Shoichet, M.S. Fast-gelling injectable blend of hyaluronan and methylcellulose for intrathecal, localized delivery to the injured spinal cord. Biomaterials 2006, 27, 2370-9.
36. Pianigiani, E.; Andreassi, A.; Taddeucci, P.; Alessandrini, C.; Fimiani, M.; Andreassi, L. A new model for studying differentiation and growth of epidermal cultures on hyaluronan-based carrier. Biomaterials 1999, 20, 1689-94.
37. Bhat, S.;Kumar, A. Cell proliferation on three-dimensional chitosan-agarose-gelatin cryogel scaffolds for tissue engineering applications. J Biosci Bioeng 2012, 114, 663-70.
38. Pooyan, P.; Tannenbaum, R.; Garmestani, H. Mechanical behavior of a cellulose-reinforced scaffold in vascular tissue engineering. J Mech Behav Biomed Mater 2012, 7, 50-9.
39. Filion, T.M.; Kutikov, A.; Song, J. Chemically modified cellulose fibrous meshes for use as tissue engineering scaffolds. Bioorg Med Chem Lett 2011, 21, 5067-70.
40. Guo, Y.; Yuan, T.; Xiao, Z.; Tang, P.; Xiao, Y.; Fan, Y.; Zhang, X. Hydrogels of collagen/chondroitin sulfate/hyaluronan interpenetrating polymer network for cartilage tissue engineering. J Mater Sci Mater Med 2012, 23, 2267-79.
41. Huang, B.; Li, C.Q.; Zhou, Y.; Luo, G.; Zhang, C.Z. Collagen II/hyaluronan/chondroitin-6-sulfate tri-copolymer scaffold for nucleus pulposus tissue engineering. J Biomed Mater Res B Appl Biomater 2010, 92, 322-31.
42. Wang, W.; Zhang, M.; Lu, W.; Zhang, X.; Ma, D.; Rong, X.; Yu, C.; Jin, Y. Cross-linked collagen-chondroitin sulfate-hyaluronic acid imitating extracellular matrix as scaffold for dermal tissue engineering. Tissue Eng Part C Methods 2010, 16, 269-79.
43. Chang, K.Y.; Hung, L.H.; Chu, I.M.; Ko, C.S.; Lee, Y.D. The application of type II collagen and chondroitin sulfate grafted PCL porous scaffold in cartilage tissue engineering. J Biomed Mater Res A 2010, 92, 712-23.
44. Jukes, J.M.; van der Aa, L.J.; Hiemstra, C.; van Veen, T.; Dijkstra, P.J.; Zhong, Z.; Feijen, J.; van Blitterswijk, C.A.; de Boer, J. A newly developed chemically crosslinked dextran-poly(ethylene glycol) hydrogel for cartilage tissue engineering. Tissue Eng Part A 2010, 16, 565-73.
45. Hoffmann, B.; Seitz, D.; Mencke, A.; Kokott, A.; Ziegler, G. Glutaraldehyde and oxidised dextran as crosslinker reagents for

chitosan-based scaffolds for cartilage tissue engineering. J Mater Sci Mater Med 2009, 20, 1495-503.

46. Smith, A.M.; Shelton, R.M.; Perrie, Y.; Harris, J.J. An initial evaluation of gellan gum as a material for tissue engineering applications. J Biomater Appl 2007, 22, 241-54.
47. Martins, A.; Chung, S.; Pedro, A.J.; Sousa, R.A.; Marques, A.P.; Reis, R.L.; Neves, N.M. Hierarchical starch-based fibrous scaffold for bone tissue engineering applications. J Tissue Eng Regen Med 2009, 3, 37-42.
48. Gomes, M.E.; Azevedo, H.S.; Moreira, A.R.; Ella, V.; Kellomaki, M.; Reis, R.L. Starch-poly(epsilon-caprolactone) and starch-poly(lactic acid) fibre-mesh scaffolds for bone tissue engineering applications: structure, mechanical properties and degradation behaviour. J Tissue Eng Regen Med 2008, 2, 243-52.
49. Vaheri, A.; Pagano, J.S. Infectious poliovirus RNA: a sensitive method of assay. Virology 1965, 27, 434-6.
50. Leong, K.W.; Mao, H.Q.; Truong-Le, V.L.; Roy, K.; Walsh, S.M.; August, J.T. DNA-polycation nanospheres as non-viral gene delivery vehicles. J Control Release 1998, 53, 183-93.
51. Erbacher, P.; Zou, S.; Bettinger, T.; Steffan, A.M.; Remy, J.S. Chitosan-based vector/DNA complexes for gene delivery: biophysical characteristics and transfection ability. Pharm Res 1998, 15, 1332-9.
52. Huang, M.; Fong, C.W.; Khor, E.; Lim, L.Y. Transfection efficiency of chitosan vectors: Effect of polymer molecular weight and degree of deacetylation. J Control Release 2005, 106, 391-406.
53. Guang Liu, W.; De Yao, K. Chitosan and its derivatives--a promising non-viral vector for gene transfection. J Control Release 2002, 83, 1-11.
54. Li, F.; Liu, W.G.; Yao, K.D. Preparation of oxidized glucose-crosslinked N-alkylated chitosan membrane and *in vitro* studies of pH-sensitive drug delivery behaviour. Biomaterials 2002, 23, 343-7.
55. Sakurai, K.; Mizu, M.; Shinkai, S. Polysaccharide--polynucleotide complexes. 2. Complementary polynucleotide mimic behavior of the natural polysaccharide schizophyllan in the macromolecular complex with single-stranded RNA and DNA. Biomacromolecules 2001, 2, 641-50.
56. Mizu, M.; Koumoto, K.; Anada, T.; Sakurai, K.; Shinkai, S. Antisense oligonucleotides bound in the polysaccharide complex and the enhanced antisense effect due to the low hydrolysis. Biomaterials 2004, 25, 3117-23.
57. Matsumoto, T.; Numata, M.; Anada, T.; Mizu, M.; Koumoto, K.; Sakurai, K.; Nagasaki, T.; Shinkai, S. Chemically modified polysaccharide schizophyllan for antisense oligonucleotides delivery to enhance the cellular uptake efficiency. Biochim Biophys Acta 2004, 1670, 91-104.

58. Azzam, T.; Eliyahu, H.; Shapira, L.; Linial, M.; Barenholz, Y.; Domb, A.J. Polysaccharide-oligoamine based conjugates for gene delivery. Journal of Medicinal Chemistry 2002, 45, 1817-24.
59. Azzam, T.; Raskin, A.; Makovitzki, A.; Brem, H.; Vierling, P.; Lineal, M.; Domb, A.J. Cationic polysaccharides for gene delivery. Macromolecules 2002, 35, 9947-53.
60. Azzam, T.; Eliyahu, H.; Makovitzki, A.; Linial, M.; Domb, A.J. Hydrophobized dextran-spermine conjugate as potential vector for *in vitro* gene transfection. J Control Release 2004, 96, 309-23.
61. Tseng, W.C.; Tang, C.H.; Fang, T.Y. The role of dextran conjugation in transfection mediated by dextran-grafted polyethylenimine. J Gene Med 2004, 6, 895-905.
62. Takei, Y.; Maruyama, A.; Ferdous, A.; Nishimura, Y.; Kawano, S.; Ikejima, K.; Okumura, S.; Asayama, S.; Nogawa, M.; Hashimoto, M.; Makino, Y.; Kinoshita, M.; Watanabe, S.; Akaike, T.; Lemasters, J.J.; Sato, N. Targeted gene delivery to sinusoidal endothelial cells: DNA nanoassociate bearing hyaluronan-glycocalyx. Faseb J 2004, 18, 699-701.
63. Pittenger, M.F.; Mackay, A.M.; Beck, S.C.; Jaiswal, R.K.; Douglas, R.; Mosca, J.D.; Moorman, M.A.; Simonetti, D.W.; Craig, S.;Marshak, D.R. Multilineage potential of adult human mesenchymal stem cells. Science 1999, 284, 143-7.
64. Takahashi, K.; Yamanaka, S. Induction of pluripotent stem cells from mouse embryonic and adult fibroblast cultures by defined factors. Cell 2006, 126, 663-76.
65. Jo, J.; Okazaki, A.; Nagane, K.; Yamamoto, M.; Tabata, Y. Preparation of cationized polysaccharides as gene transfection carrier for bone marrow-derived mesenchymal stem cells. J Biomater Sci Polym Ed 2010, 21, 185-204.
66. Jo, J.I.; Nagaya, N.; Miyahara, Y.; Kataoka, M.; Harada-Shiba, M.; Kangawa, K.;Tabata, Y. Transplantation of Genetically Engineered Mesenchymal Stem Cells Improves Cardiac Function in Rats With Myocardial Infarction: Benefit of a Novel Nonviral Vector, Cationized Dextran. Tissue Eng 2006.
67. Gilmore, I.R.; Fox, S.P.; Hollins, A.J.; Akhtar, S. Delivery strategies for siRNA-mediated gene silencing. Curr Drug Deliv 2006, 3, 147-5.
68. Nagane, K.; Jo, J.; Tabata, Y. Promoted adipogenesis of rat mesenchymal stem cells by transfection of small interfering RNA complexed with a cationized dextran. Tissue Eng Part A 2010, 16, 21-31.
69. Hong, J.H.; Yaffe, M.B. TAZ: a beta-catenin-like molecule that regulates mesenchymal stem cell differentiation. Cell cycle (Georgetown, Tex 2006, 5, 176-9.

70. Okazaki, A.; Jo, J.I.; Tabata, Y. A Reverse Transfection Technology to Genetically Engineer Adult Stem Cells. Tissue Eng 2006,
71. Kido, Y.; Jo, J.; Tabata, Y. A gene transfection for rat mesenchymal stromal cells in biodegradable gelatin scaffolds containing cationized polysaccharides. Biomaterials 2011, 32, 919-25.
72. Weissleder, R.; Mahmood, U. Molecular imaging. Radiology 2001, 219, 316-33.
73. Jokerst, J.V.; Gambhir, S.S. Molecular imaging with theranostic nanoparticles. Acc Chem Res 2011, 44, 1050-60.
74. Villaraza, A.J.; Bumb, A.; Brechbiel, M.W. Macromolecules, dendrimers, and nanomaterials in magnetic resonance imaging: the interplay between size, function, and pharmacokinetics. Chem Rev 2010, 110, 2921-59.
75. Cromer Berman, S.M.; Walczak, P.;Bulte, J.W. Tracking stem cells using magnetic nanoparticles. Wiley Interdiscip Rev Nanomed Nanobiotechnol 2011, 3, 343-55.
76. Bhirde, A.; Xie, J.; Swierczewska, M.; Chen, X. Nanoparticles for cell labeling. Nanoscale 2011, 3, 142-53.
77. Zhang, C.; Liu, T.; Gao, J.; Su, Y.; Shi, C. Recent development and application of magnetic nanoparticles for cell labeling and imaging. Mini Rev Med Chem 2010, 10, 193-202.
78. Tartaj, P.; Morales, M.D.; Veintemillas-Verdaguer, S.; Gonzalez-Carreno, T.; Serna, C.J. The preparation of magnetic nanoparticles for applications in biomedicine. Journal of Physics D-Applied Physics 2003, 36, R182-R97.
79. Cheng, F.Y.; Su, C.H.; Yang, Y.S.; Yeh, C.S.; Tsai, C.Y.; Wu, C.L.; Wu, M.T.; Shieh, D.B. Characterization of aqueous dispersions of Fe(3)O(4) nanoparticles and their biomedical applications. Biomaterials 2005, 26, 729-38.
80. Jo, J.; Aoki, I.; Tabata, Y. Design of iron oxide nanoparticles with different sizes and surface charges for simple and efficient labeling of mesenchymal stem cells. J Control Release 2010, 142, 465-73.
81. Cheng, Z.; Al Zaki, A.; Hui, J.Z.; Muzykantov, V.R.; Tsourkas, A. Multifunctional nanoparticles: cost versus benefit of adding targeting and imaging capabilities. Science 2012, 338, 903-10.
82. Cheng, Y.; Zhao, L.; Li, Y.; Xu, T. Design of biocompatible dendrimers for cancer diagnosis and therapy: current status and future perspectives. Chem Soc Rev 2011, 40, 2673-703.
83. Saravanakumar, G.; Kim, K.; Park, J.H.; Rhee, K.; Kwon, I.C. Current status of nanoparticle-based imaging agents for early diagnosis of cancer and atherosclerosis. J Biomed Nanotechnol 2009, 5, 20-35.

84. Tang, T.Y.; Muller, K.H.; Graves, M.J.; Li, Z.Y.; Walsh, S.R.; Young, V.; Sadat, U.; Howarth, S.P.; Gillard, J.H. Iron oxide particles for atheroma imaging. Arterioscler Thromb Vasc Biol 2009, 29, 1001-8.

85. Meoli, D.F.; Sadeghi, M.M.; Krassilnikova, S.; Bourke, B.N.; Giordano, F.J.; Dione, D.P.; Su, H.; Edwards, D.S.; Liu, S.; Harris, T.D.; Madri, J.A.; Zaret, B.L.; Sinusas, A.J. Noninvasive imaging of myocardial angiogenesis following experimental myocardial infarction. J Clin Invest 2004, 113, 1684-91.

86. Leong-Poi, H.; Christiansen, J.; Heppner, P.; Lewis, C.W.; Klibanov, A.L.; Kaul, S.; Lindner, J.R. Assessment of endogenous and therapeutic arteriogenesis by contrast ultrasound molecular imaging of integrin expression. Circulation 2005, 111, 3248-54.

87. Hua, J.; Dobrucki, L.W.; Sadeghi, M.M.; Zhang, J.; Bourke, B.N.; Cavaliere, P.; Song, J.; Chow, C.; Jahanshad, N.; van Royen, N.; Buschmann, I.; Madri, J.A.; Mendiwzabal, M.;Sinusas, A.J. Noninvasive imaging of angiogenesis with a 99mTc-labeled peptide targeted at alphavbeta3 integrin after murine hindlimb ischemia. Circulation 2005, 111, 3255-60.

88. Winter, P.M.; Caruthers, S.D.; Allen, J.S.; Cai, K.; Williams, T.A.; Lanza, G.M.; Wickline, S.A. Molecular imaging of angiogenic therapy in peripheral vascular disease with alphanubeta3-integrin-targeted nanoparticles. Magn Reson Med 2010, 64, 369-76.

89. Dvorak, H.F. Angiogenesis: update 2005. J Thromb Haemost 2005, 3, 1835-42.

90. Dobrucki, L.W.; Sinusas, A.J. Imaging angiogenesis. Curr Opin Biotechnol 2007, 18, 90-6.

91. Marui, A.; Tabata, Y.; Kojima, S.; Yamamoto, M.; Tambara, K.; Nishina, T.; Saji, Y.; Inui, K.; Hashida, T.; Yokoyama, S.; Onodera, R.; Ikeda, T.; Fukushima, M.; Komeda, M. A novel approach to therapeutic angiogenesis for patients with critical limb ischemia by sustained release of basic fibroblast growth factor using biodegradable gelatin hydrogel: an initial report of the phase I-IIa study. Circ J 2007, 71, 1181-6.

92. Jo, J.; Lin, X.; Nakahara, T.; Aoki, I.; Saga, T.; Tabata, Y. Preparation of polymer-based magnetic resonance imaging contrast agent to visualize therapeutic angiogenesis. Tissue Eng Part A 2013, 19, 30-9.

93. Massoud, T.F.; Gambhir, S.S. Molecular imaging in living subjects: seeing fundamental biological processes in a new light. Genes Dev 2003, 17, 545-80.

94. Kozloff, K.M.; Weissleder, R.; Mahmood, U. Noninvasive optical detection of bone mineral. J Bone Miner Res 2007, 22, 1208-16.

95. Kempen, D.H.; Yaszemski, M.J.; Heijink, A.; Hefferan, T.E.; Creemers, L.B.; Britson, J.; Maran, A.; Classic, K.L.; Dhert, W.J.; Lu, L. Non-invasive monitoring of BMP-2 retention and bone formation in composites for bone tissue engineering using SPECT/CT and scintillation probes. J Control Release 2009, 134, 169-76.

96. Liu, J.; Jo, J.; Kawai, Y.; Aoki, I.; Tanaka, C.; Yamamoto, M.; Tabata, Y. Preparation of polymer-based multimodal imaging agent to visualize the process of bone regeneration. J Control Release 2012, 157, 398-405.

97. Yamamoto, M.; Takahashi, Y.; Tabata, Y. Enhanced bone regeneration at a segmental bone defect by controlled release of bone morphogenetic protein-2 from a biodegradable hydrogel. Tissue Eng 2006, 12, 1305-11.

2 Starch based Polymers for Drug Delivery Applications

Fernando G. Torres* **and Omar P. Troncoso**

Polymers and Bio nanomaterials Laboratory, Pontificia Universidad Catolica del Peru, Lima 32 – Peru.

2.1 Drug Delivery Systems

2.1.1 Conventional Free Drug Administration

Conventional administration of drugs presents several disadvantages. For instance, Allen and Cullis[1] have listed common problems and non-ideal properties of drugs administrated parentally. These problems include poor solubility, tissue damage on extravasation, rapid breakdown of the drug *in vivo*, unfavourable pharmacokinetics, poor biodistribution and lack of selectivity of target tissues.

Poor solubility can hinder the achievement of a convenient pharmaceutical format as hydrophobic drugs can precipitate in aqueous media, whereas unfavourable pharmacokinetics impliy that the drug is cleared too rapidly from the body, requiring high doses or continuous infusion. The fact that drugs can widespread throughout the body, affecting normal tissues, might limit the dose used[1].

In contrast, polymer based drug delivery systems (DDS) can be designed to alter the pharmacokinetics and the duration of action of drugs. For instance, Burnham[2] has covalently coupled polymers, such as polyethylene glycol (PEG), to proteins, such as interferon, in order to make them last up to one week in humans. Other studies have used polymer-drug conjugates in cancer chemotherapy. This alters the biodistribution of anticancer drugs as the polymer-drug conjugate accumulates more in tumours which have a leaky vascular bed[3-5]. Other systems are designed to overcome tissue barriers such as lung, skin, intestine, etc [6].

* ***Corresponding Author***

According to Langer and Peppas[6], controlled drug delivery systems use one of the following mechanisms to control de release of drugs:

1. *Diffusion:* Polymeric reservoirs (drugs are surrounded by a polymeric membrane) or polymeric matrix (drugs are distributed through the polymer) release drugs. The diffusion through the polymer is the rate-limited step.
2. *Chemical reaction:* The polymer must be degraded by water or a chemical reaction in order to release the drug.
3. *Solvent activation and transport:* The drug is released by the swelling of the polymer in which the drug was previously locked into place within the polymer matrix.

2.1.2 Drug Delivery Systems Classification

Drug delivery systems can be classified according to different criteria. One criterion can be the route used by the DDS. The different routes used include: intravenous, nasal, transdermal, pulmonary, buccal, ocular, vaginal, rectal and oral delivery routes[7].

Many DDS are environmentally responsive. Typical stimuli include temperature, pH, electric field, light, magnetic field and concentration of electrolytes and glucose[8]. Table 2.1 lists several stimuli responsive DDS.

In this paper we review the different starch based DDS. We have classified these systems into the following categories:

1. Gels and Hydrogels
2. Micelles
3. Capsules
4. Microparticles
5. Nanoparticles

TABLE 2.1

List of main stimuli responsive drug delivery systems

Stimuli	Study	Reference
Temperature	Poly(N-isopropylacrylamide): experiment, theory and application.	Schild,1992
	Thermosensitive biodegradable hydrogels for the delivery of therapeutic agents.	Kwon, 2005
	Synthesis and characterization of thermoresponsive graft copolymers of NIPAAmand 2-alkyl-2-oxazolines by the "grafting from" method.	Rueda, 2005

TABLE 2.1 ***Contd...***

Stimuli	Study	Reference
pH	pH-sensitive polymers for drug delivery.	Na, 2005
	Polymeric micelle for tumor pH and folate-mediated targeting.	Lee,2003
	Dynamic swelling behavior of pH-sensitive anionic hydrogels used for protein delivery.	Kim, 2003
	Characterising the size and shape of polyamidoamines in solution as a function of pH using neutron scattering and pulsed-gradient spin-echo NMR.	Khayat, 2006
Electric field	Interaction of hydrophobically-modified poly-N-isopropylacrylamides with model membranes – or playing a molecular accordion.	Ringsdorf, 1991
	Effect of molecular weight and polydispersity on kinetics of dissolution and release from pH/temperature-sensitive polymers.	Ramkissoon-Ganorkar, 1999
	Genetic engineering of protein-based polymers: the example of elastinlike polymers.	Rodriguez-Cabello, 2006
agnetic field	Evaluation of an elastin-like polypeptide-doxorubicin conjugate for cancer therapy.	Dreher,2003
	Structural optimization of a "smart" doxorubicin-polypeptide conjugate for thermally targeted delivery to solid tumors.	Furgeson, 2006
	A new concept for macromolecular therapeutics in cancer-chemotherapy—mechanism of tumoritropic accumulation of proteins and the antitumor agent SMANCS.	Matsumura, 1986
Others	Poly(amidoamine)s as potential nonviral vectors: ability to form interpolyelectrolyte complexes and to mediate transfection *in vitro*.	Richardson, 2001

2.1.3 Polymeric Materials in Drug Delivery

Polymers are used for clinical applications due to the fact that their wide architecture (linear, branched, cross-linked, block, graft, multivalent, dendronized, etc.) offer diversity in chemistry, dimensions and topology[23]. Other important factors to take into account, when using polymers in drug delivery, are chemical composition (polyester, polyanhydride, polyamide), backbone stability (biodegradable, non-biodegradable); and water solubility (hydrophilic, hydrophobic)[23].

A variety of polymeric materials are used for drug delivery systems, including synthetic polymers (poly(ethylene glycol), N-(2-hydroxypropyl) methacrylamide co-polymers, poly(vinylpyrrolidone), poly(ethyleneimine) and linear polyamidoamines); natural polymers (dextran (a-1,6 polyglucose), dextrin (a-1,4 polyglucose), hyaluronic acid and chitosans); and pseudosynthetic polymers (the man-made poly(amino acids) poly(L-lysine), poly(L-glutamic acid), poly(malic acid) and poly(aspartamides))[24].

Drugs can be physically entrapped within polymer shells and matrices, or they can be covalently attached to the polymer backbone. According to Duncan and Spreafico[25] polymer–drug conjugates alter the pharmacokinetics of the drug by increasing the drug's effective molecular weight.

However, there are some concerns regarding the use of polymers for drug delivery systems. The polydispersity of polymers is of particular importance. Drugs are homogeneous and defined species while polymers are inherently heterogeneous mixtures of chains of varying lengths. The suitability of polydisperse polymers must be studied as biological properties are molecular weight dependent[26]. Another important factor is biocompatibility as polymers are often internalized by cells. Polycations are often cytotoxic, haemolytic and complement-activating, whereas polyanions are less cytotoxic but can induce anticoagulant activity and cytokine release[27].

2.2 Starch as a Biomaterial

2.2.1 Native Starch

Starch can be extracted from roots, seeds, stems and tubers of different plants, such as corn, potato, wheat, rice and others. It is formed by granules that are composed by two different polymers, namely amylose and amylopectin. Amylose is linear, comprises 20-30% of the starch granule and is formed by long chains of α-(1,4)-linked D-glucose units with a degree of polymerization ranging 3×102 - 1×104[28]. In contrast, amylopectin is a branched polymer and comprises 70-80% of the granule. It has α-(1,4)-linked glucose chains, joined by -(1,6)-linkages with a degree of polymerisation of about 108[28].

With the right amount of water and heat, starch can be processed as a thermoplastic and can be used to produce a variety of products such as films, foams, sheets, etc. Multiple chemical and physical reactions; such as water diffusion, granular expansion, gelatinization, decomposition, melting and crystallization; may occur during processing. Among these phase transitions, gelatinization is particularly important because it is closely related to the others, and it is the basis of the conversion of starch thermoplastic[29].

According to Marques et al.[30] biocompatibility is an inherent property of structures derived from organic polymers like starch based materials. The biocompatibility shown by starch based products is due to the presence of major biocompatible structural components as starch polymer molecules and by-products obtained from partial hydrolysis.

In general, starch-based materials exhibit attributes of biocompatibility and have been used in several biomedical applications[31-36]. Studies carried out by Torres et al.[37] have confirmed the biocompatibility of native starch products. They have used starch from 17 different Andean crops in order to prepare films for cell seeding. Their results confirmed the *in vitro* biocompatibility of starch films with 3T3 fibroblast cells. Figure 2.1 shows the proliferation of 3T3 fibroblast cell lines after three days in a control flask and in an Andean potato starch film. The cells showed similar morphology among the different types of films and the control group.

In spite of its biocompatibility, native starch based products have some disadvantages that limit its use in a variety of applications. Such disadvantages include poor mechanical properties and the highly hydrophilic character of starch based products. Hence, modified versions of native starch have been produced.

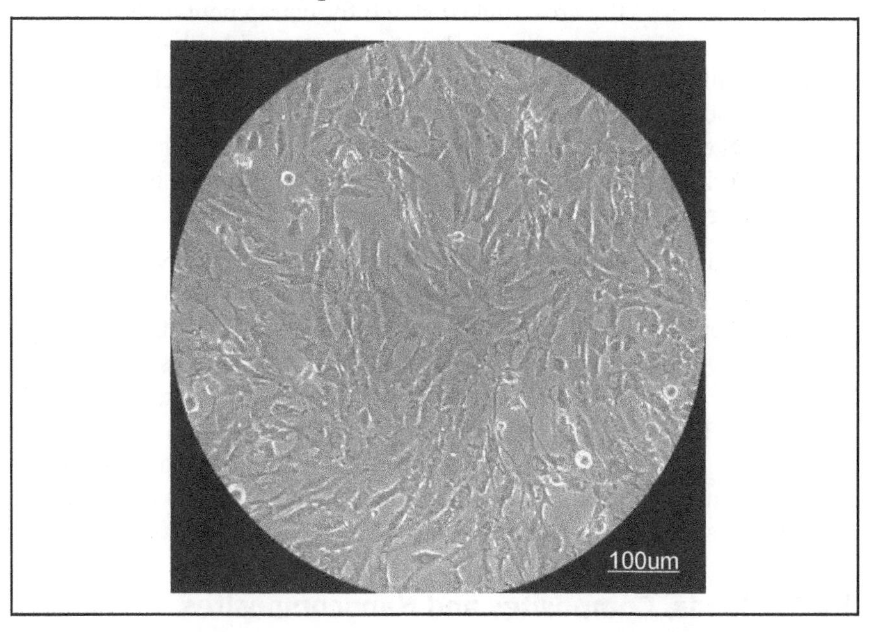

FIGURE 2.1 Proliferation of 3T3 fibroblast cell line after three days in a potato starch film. Adapted from Torres (2011).

2.2.2 Modified Starch

Modification treatments of starch in the food industry include gelatinization, monoester or cross-linked phosphorylation, hydroxy-propylation, etc[38-39]. Modified starches for biomedical applications include cadexomer iodine, oxidized starch and hydroxyethyl starch.

Cadexomer is a three dimensional hydrophilic starch crosslinked with epichlorohydrin incorporated with 0.9% (w/w) iodine[40]. Oxidized starch is produced by the reaction of starch and an oxidizing agent under a controlled temperature and pH. It is used due to its low viscosity, high stability, film-forming and binding properties[41,42]. Hydroxyethyl starch (HES) is obtained through hydrolysis and subsequent hydroxyethylation of amylopectin. Low substitute HES is usually produced by the reaction of starch with ethylene oxide at temperatures below 50 °C in aqueous slurries in the presence of a swelling inhibiting salt[43].

Modification treatments applied to native starches for the production of drug delivery systems include self-association (induced by changes in pH, ionic strength or physical and thermal means), complexation with salts and covalent crosslinking[44-46].

Kost and Shefer[47] ionically cross-linked corn starch with calcium chloride. They used the cross-linked starch for entrapment and controlled release of bioactive molecules such as salicyclic acid. Shalviri et al.[48] prepared hydrogels by cross-linking starch with varying levels of xanthan gum and sodium trimetaphosphate (STMP). Their results suggest that such hydrogels can be potentially used as a film-forming material in controlled release formulations. Reis et al.[49] also prepared hydrogels by means of a cross-linking polymerization technique.

Starch graft copolymers have also been used as drug delivery systems. Simi and Abraham[50] reported the grafting of fatty acid on starch using potassium persulphate as catalyst. They used the modified starch in order to produce starch nanoparticles loaded with indomethacin, as model drug. Shaikh et al.[51] reported the preparation of acrylic monomers-starch graft copolymers by means of a ceric ion initiation method. They used paracetamol as a model drug. Their results indicate that the graft copolymers may be useful to overcome the stomach's harsh environment and can be used as excipients in colon-targeting matrices.

2.2.3 Blends, Composites and Nanocomposites

Starch based blends, composites and nanocomposites are produced in order to overcome some of the disadvantages of native starch. Polymer blending is a well-used technique whenever modification of properties is required, because it uses conventional technology at low cost[52]. Several studies report the blend of starch with various polyolefins[53-56]. However, these starch blends are not biodegradable and cannot be used as biomaterials.

Blends of starch with other natural materials have also been studied. Warth et al.[57] have developed starch/cellulose acetate blends. Arvanitoyannis and Biliaderis[58] studied aqueous blends of methyl cellulose and soluble starch, plasticized with glycerol or sugars, prepared by casting or by extrusion and hot pressing. Pereira et al.[59] reported the production of biodegradable hydrogels based on corn starch/cellulose acetate blends, produced by free-radical polymerization with methyl methacrylate or an acrylic acid monomer.

Starch blends have also been used in the production of drug delivery systems[60-63]. Soares et al.[62] have studied blends of cross-linked high amylose starch and pectin loaded with diclofenac. The rheological properties of the samples demonstrated that drug loading resulted in the formation of weaker gels whereas the increase of pectin ratio contributes to originate stronger structures.

Silva et al.[63] have studied physical blends of starch graft copolymers as matrices for colon targeting drug delivery systems. They combined high amylose starch with two kinds of acrylic monomers (methacrylic acid and 2-hydroxyethyl methacrylate). Each of the monomers offers different properties such as permeability for water and drugs, pH sensitivity and biodegradability. It was observed that these physical blend matrices offer good controlled release of drugs, as well as of proteins and present suitable properties to be used as hydrophilic matrices for colon-specific drug delivery systems.

Starch has also been used as matrix for the preparation of composites and nanocomposites. For instance, Torres et al.[64,65] have prepared biodegradable composites of starch reinforced with lignocellulosic fibres. These composites show improved mechanical properties in comparison to natural starch materials. Marques et al.[66] have evaluated the cytotoxicity and cell adhesion, and studied the biocompatibility of starch-hydroxyapatite composites by means of *in vitro* studies. Saikia et al.[67] have studied biocompatible starch/polyaniline composites. The hemolysis prevention activity of such composites is found to increase, as compared to the pure polyaniline, with minor compromise in the antioxidant activity.

Starch nanocomposites have also been studied. Torres et al.[68] have used cellulose nanofibers in order to produce starch-cellulose nanocomposites. Chen et al.[68] have prepared starch-clay composites with various types of clay, including smectite clays, montmorillonite and hectorite. Their results show that the presence of clay increased the elastic modulus of starch in all cases.

2.3 Starch based Drug Delivery Systems

2.3.1 Gels and Hydrogels

According to Lee[70] gels are polymer networks which swell when immersed in solvent. They are prevented from dissolving by the presence of crosslinks which hold their structure intact. For instance, Yoon et al.[71] have used retrograded waxy maize starch gels as tablet matrix for controlled release of theophylline. They found that temperature-cycled retrogradation of waxy maize starch gel provided a compact matrix structure that effectively retarded the release of theophylline. Microgels suspensions have also been used as drug delivery systems Li et al.[72-74] used oxidated starch to prepare starch microgel particles (10-20 um) that are able to uptake and release lysozyme. The uptake and release kinetics have been found to be related to the pH of the medium.

Hydrogels are gels that are able to absorb large amounts of water. Several applications have been found for hydrogels as they can exhibit water-, thermo- or pH-sensibility. However, starch based hydrogels exhibit some disadvantages such as poor mechanical strength, which limits their application. Thus, investigations mainly report the use of starch together with other gelling forming components, including xanthan gum metacrylic acic[75], polyacrylic acid[76], ethylene-co-vinil alcohol[77], cellulose acetate[59], etc.

Several works have been published using pH-responsive hydrogels for drug delivery systems. For instance, Ali et al.[75] have studied the swelling kinetics of starch/methacrylic acid hydrogels. Such hydrogels showed a Fickinan diffusion behaviour at pH 1 and non-Fickian diffusion at pH 7. Thus, ketoprofen loaded samples keep the loaded drug at pH 1 and release it at pH 7. Figure 2.2 shows the pH-dependent swelling behaviour of starch/ methacrylic acid hydrogels of this study. Samples exhibit a pH-dependent phase transition, with higher swelling degrees displayed at high pH values ($pH > 4$).

In order to determine the nature of diffusion of water into hydrogels, the following equation can be used[78]:

$$\frac{M_t}{M_\infty} = Kt^n \qquad .....(1)$$

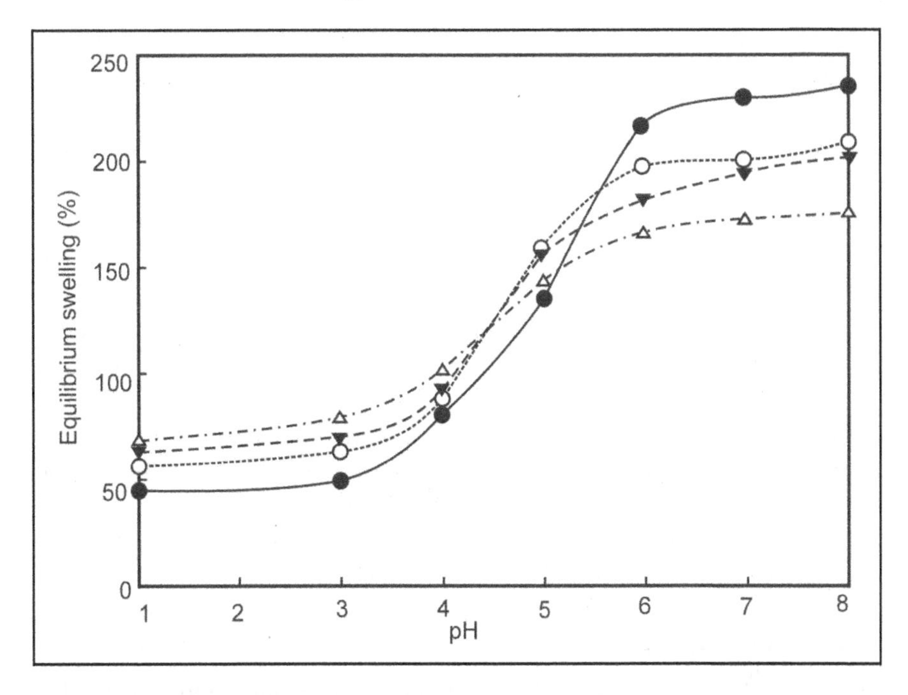

FIGURE 2.2 pH-dependent swelling of starch/methacrylic acid copolymer hydrogels at different starch content (wt.%). (•) 5, (○) 10, (▾) 15 and (Δ) 20. Extracted from Ali (2009).

where M_t and M_∞ stand for the amount of solvent diffused into the gel at time t and at infinite time respectively, K is a constant related to the structure of the network and n is a characteristic exponent of the transport mode of the solvent.

The value of the characteristic exponent is used to define three classes of diffusion that depend on the relative rates of diffusion and polymer[79]:

- *Case I or Fickian diffusion:* The rate of diffusion is much less than that of relaxation ($n = 0.50$)
- *Case II diffusion:* Diffusion is very rapid compared with the relaxation processes ($n = 1$). This case corresponds to the most desirable kinetic behaviour for controlled release.

- *Non-Fickian diffusion:* Diffusion and relaxation rates are comparables ($0.50 < n < 1$).

By controlling the diffusion mechanism, an on-off switch behaviour can be achieved for hydrogels. Bajpai et al.[76] studied starch/poly(acrylic acid) hydrogels. The samples underwent a sharp transition from Fickian swelling behaviour to non-Fickian swelling behaviour as the pH of the swelling medium changed from 2.0 to 7.4. Also, Sadegi et al. [80] prepared starch/poly(sodium acrylate-co-acrylamide) hydrogels loaded with Ibuprofene, a poor water soluble drug. Simulated gastric and intestinal pH conditions were used to evaluate Ibuprofene release. The results showed that the release was much quicker at pH 7.4 than at pH 1.2. Pereira et al.[59] starch/cellulose acetate hydrogels also showed a similar pH sensitive behaviour.

Other starch-based hydrogels are not affected by pH. Reis et al.[49] (2008) have used chemically modified starch to prepare drug delivery hydrogels. The swelling behaviour of their samples was not significantly affected by changes of pH or temperature. Other authors have used starch as polymeric binders. Shaaheen et al.[81] have studied the effect of starch on terbutaline hydrogels of PVA/NaCl/H_2O. They found that the inclusion of starch increased the degree of swelling compared with the hydrogels that used other binders such as gelatin.

2.3.2 Micelles

Micelles are collections of amphiphilic surfactant molecules that spontaneously aggregate water into a spherical vesicle. The centre of the micelles is hydrophobic and can storage hydrophobic drugs until they are released by some drug delivery mechanism Husseini[82]. Modified starch has been used to prepare micelles. For instance, Besheer et al.[83] used hydroxyethyl starch (HES), a water soluble semisynthetic polysaccharide used in medicine as a plasma volume expander. They esterified HES with lauric, palmitic and stearic acids in order to produce 20-30 nm micelles.

Zhang et al.[84] have used disulfide core-crosslinked micelles based on amphiphilic starch and graft-poly(ethylene glycol). These micelles are responsive to the presence of glutathione (GHS), a tripeptide produced in the human body. The micelles used in this study were loaded with a model anticancer drug doxorubicin (DOX). The results showed that only a small amount of DOX was released in PBS solution without GSH, while a quick release occurred in the presence of 10 mM GHS.

Ju et al.[85] have reported the use of a starch based thermoresponsive polymer, namely 2-hydroxy-3-butoxypropyll starch (HBPS). HBPS was prepared changing the hydrophobic-hydrophilic balance of starch using butyl glycidyl ether as hydrophobic reagent. In water, HBPS self-assembles into micelles below the lower critical solution temperature (the temperature below which the components are miscible). Ju et al.[85] found that drug loaded HBPS micelles are thermoresponsive as the drug release is accelerated above the critical solution temperature.

2.3.3 Capsules

Encapsulation was first proposed by Chang[86] who envisaged the use of ultrathin polymer membrane microcapsules for the immunoprotection of transplanted cells. Then, Lim et al.[87] reported that encapsulated pancreatic cells corrected the diabetic state of rats with streptozotocin-induced diabetes for several weeks.

Encapsulation technology has been used in medicine, pharmaceutics, agriculture, and the cosmetic industries for the development of controlled-release delivery systems[88,89]. Capsules for drug delivery are systems in which the drug is generally confined to a cavity consisting of an inner liquid core surrounded by a polymeric membrane. In this case the active substances are usually dissolved in the inner core but may also be adsorbed to the capsule surface.

Several authors have studied the use of starch based microspheres for the encapsulation of drugs. Modified soluble starch is often used together with a crosslinking agent through an emulsification technique[90-93,46] in order to obtain starch based microspheres. Figure 2.3 shows an example of crosslinked starch microspheres prepared by a water-in-water emulsion method. The loading and release kinetics depend on a variety of factors. Fang et al.[92] have found that the release profile of crosslinked starch microspheres loaded with Methylene Blue (MB) is influenced by loading time, dissolution medium, loading temperature and MB concentration.

As in the case of other starch based drug delivery systems, different strategies are followed in order to control the release of drugs. For instance, the microspheres prepared by Malafaya et al.[46] seem to be pH-dependent. The results published by Mundargi et al.[93] show that the microspheres prepared by an emulsification technique display a diffusion coefficient decreases with croslinking and increases with the starch content. They were able to use their microspheres to prepare Ampicillin tablets that were effective in realizing the drug over an extended period of about 24h. Fundueanu et al.[94] have prepared thermo-sensitive starch based microspheres for temperature-controlled release of drugs. They used modified starch that poses strong anionic functional groups (-SO_3H) capable of binding electrostatically to drugs.

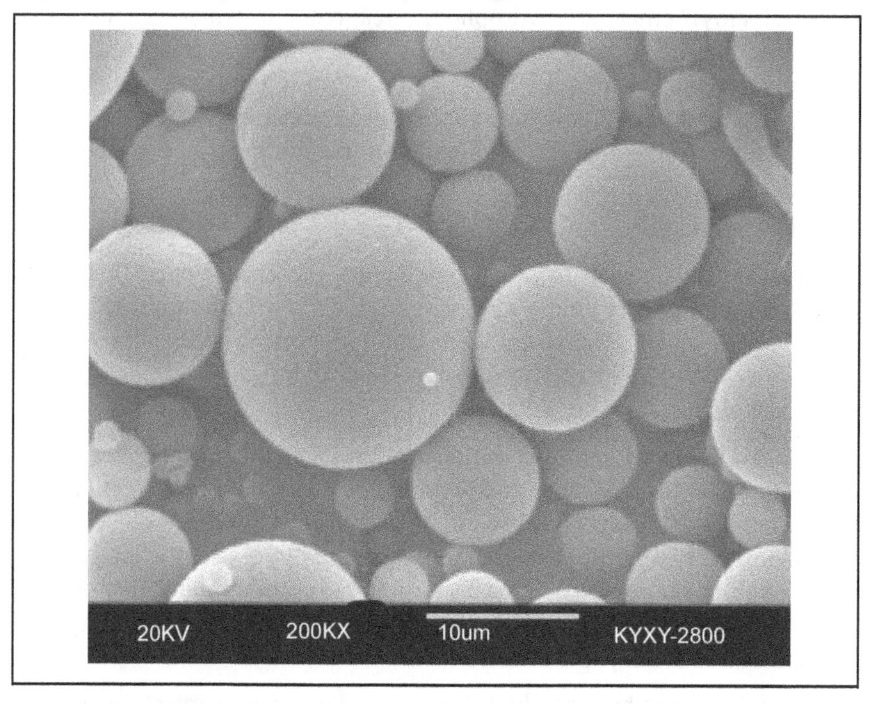

FIGURE 2.3 SEM photographs (2000×) of crosslinked starch microspheres. Adapted from Li (2012).

2.3.4 Microparticles

Microparticles are particles in the micrometre size range. These particles can be used as drug carriers. Drugs are incorporated in the form of solid solutions or solid dispersions, or they are adsorbed or chemically bounded. For instance, Baimark et al. (Y; Srisa-ard, M; Srihanam, P) have studied the morphology and thermal stability of silk fibroin/starch blended microparticles. Biodegradable microparticles of silk fibroin (SF)/starch blends were prepared by a simple water-in-oil emulsion solvent diffusion technique. The influence of SF/starch ratios on characteristics of the blended microparticles was investigated. The results suggested that SF conformational transition, thermal stability, morphology and dissolution of the blended microparticles can be adjusted by varying the blended ratio.

Several authors have investigated the applications of starch microparticles in drug delivery and tissue engineering (TE). Tuovinen et al.[96] have studied drug release from starch-acetate microparticles and films with and without incorporated alpha-amylase. Balmayor et al. [97] demonstrated the successful preparation of starch-poly-epsilon-caprolactone microparticles incorporating dexamethasone (DEX). These microparticular systems seem to be quite promising for controlled release applications, namely as carriers of important differentiation agents in TE. In the case of the controlled delivery of diclofenac sodium, Desai, KG[21] has studied the properties of tableted high-amylose corn starch-pectin blend microparticles. This study indicates that tableted microparticulate system is found to be suitable for the manipulation of release behavior for DS in the gastrointestinal tract.

2.3.5 Nanoparticles

Nanoparticles have been studied due to their physical, mechanical, magnetic, chemical and biological applications. Recent studies show that nanoparticles have the ability to protect drugs from degradation in the gastrointestinal tract and can deliver poorly water soluble drugs [99]. Panyam and Labhasetwar[100] have reported that the uptake of nanostructures is 15-250 times greater than that of microparticles in the 1-10um range. According to Hughes[101] nanotechnology can improve the performance and acceptability of dosage forms due to the fact that nanoparticles are able to penetrate tissues and are easily taken up by cells.

However, the safety of the use of nanoparticles in humans is still to be determined. Guzman et al.[102] have reported that copper, titanium and silicon and their oxides nanoparticles have induced inflammatory and toxic effects on cells. The use of nanoparticles of biodegradable materials could avoid undesired toxic effects.

Shalviri et al.[103] have studied the design of pH-responsive nanoparticles for controlled delivery of anticancer drug doxorubicin (Dox) using nanoparticles of poly (methacrylic acid)-polysorbate 80-grafted starch (PMAA-PS 80-g-St). Figure 2.4 shows a TEM image of the nanoparticles prepared by Shalviri et al. They confirmed that the chemical composition of the graft polymer and exhibited pH-dependent swelling in a physiological pH range. The results suggest that the new pH-responsive polymer nanoparticles are useful in controlled drug delivery.

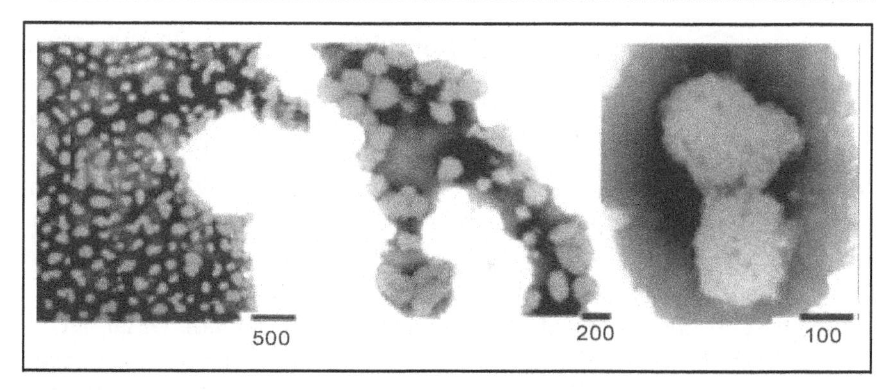

FIGURE 2.4 TEM images of nanoparticles of terpolymer of poly(methacrylic acid), polysorbate 80 and starch. Adapted from Shalviri (2013).

Shi et al.[104] have investigated the rheological properties of suspensions containing cross-linked starch nanoparticles prepared by spray and vacuum freeze drying. The suspensions containing 10% (w/w) spray dried and vacuum freeze dried nanoparticles. They compared the suspensions containing vacuum freeze dried nanoparticles to have more shear thinning and less thixotropic behavior with to those containing spray dried Nanoparticles.

Dandekar et al.[105] have used hydrophobic Starch nanoparticles for the delivery of docetaxel. They used a hydrophobic starch derivative for safe and enhanced delivery of anticancer agents effective against numerous types of cancers. Uptake studies with these nanoparticles indicate their enhanced internalization by the cancerous cells and their peri-nuclear localization.

2.3.6 Other Starch based Drug Delivery Systems

Other type of starch based drug delivery systems includes the use of different structures such as films and fibres. For instance, Arockianathan[106] have evaluated the use of films of alginate and sago starch, impregnated with silver nanoparticles with antibiotic gentamicin prepared by solvent casting method. They found that these films might be a potential and economical wound dressing material.

Pea starch based film coatings have been used for site-specific drug delivery to the colon of inflammatory bowel disease patients. It has been found that such coatings are poorly permeable in media simulating the contents of the stomach and small intestine but once the colon is reached, drug release sets on and is time controlled. This can be attributed to the partial degradation of the peas starch by enzymes secreted by bacteria, which are preferentially present in the colon Karrout[107].

Wang et al.[108] prepared fibres of alginate and starch, with salicylic acid as model drug by spinning their solution through a viscose-type spinneret into a coagulating bath containing aqueous $CaCl_2$ and ethanol. Their results showed that the amount of salicylic acid released increased with an increase in the proportion of starch present in the fibre. The alginate/starch fibres were also sensitive to pH and ionic strength.

Conclusion

In this paper we have reviewed the use of starch in drug delivery systems. In most cases, starch is not used in its native state but has undergone modification treatments such as self-association, complexation with salts and covalent crosslinking. Other approaches include the use of starch to prepare blends and composites. The drug delivery systems reviewed include gels and hydrogels, micelles, capsules, microparticles and nanoparticles. Most of such systems are environmentally responsive. Typical stimuli include temperature, pH, electric field, light, magnetic field and concentration of electrolytes and glucose. Starch is a biomaterial with different applications in the biomedical field. Future prospects include the use of starch to produce amylopectin nanocrystals that could be used in different biomedical application such as wound dressings, scaffolds for tissue engineering and drug delivery systems.

References

1. Allen T.M. and Cullis P.R., (2004), Drug Delivery Systems: Entering the Mainstream, Science, 19, pp. 1818-1822.
2. Burnham, N.L., (1994), Polymers for delivering peptides and proteins, American Journal of Health-System Pharmacy, 51(2), pp. 210-218.
3. Duncan R., Dimitrijevic, S. and Evagorou E.G., (1996), The role of polymer conjugates in the diagnosis and treatment of cancer, S.T.P. Pharma. Sci., 6, pp. 237-263.
4. Putnam, D., and Kopecek, J., (1995), Polymer conjugates with Anticancer Activity, Advances in Polymer Science, 122, pp 55-123.
5. Murakami Y., Tabata, Y. and Ikada, Y., (1997), Tumor accumulation of poly(ethylene glycol) with different molecular weights after intravenous injection, Drug Delivery, 4(1), pp 23-31.

6. Langer, R. and Peppas, N.A., (2004), Advances in biomaterials, drug delivery, and bionanotechnology, Bioengineering, Food and Natural Products, 49(12), pp. 2990-3006.
7. Lehr, C.M, (1994), Bioadhesion technologies for the delivery of peptide and protein drugs to the gastrointestinal tract, Critical Reviews in Therapeutic Drug Carrier Systems, 11(2-3), 119-160.
8. Zoccola, M., Aluigi, A. and Tonin, C., (2009), Characterisation of keratin biomass from butchery and wool industry wastes, Journal of Molecular Structure, 938 (1-3), pp. 35-40.
9. Schild, H.G., (1992), Poly(N-isopropylacrylamide): experiment, theory and application, Progress in Polymer Science, 17(2), pp. 163-249.
10. Kwon, Y.M. and Kim, S.W., (2005), Polymeric Drug Delivery Systems, Taylor & Francis, Boca Raton, pp. 251-274.
11. Rueda, J.C., Zschoche, St., Komber, H., Schmaljohann, D. and Voit, B., (2005), Synthesis and Characterization of Thermore-sponsive Graft Copolymers of NIPAAm and 2-Alkyl-2-oxazolines by the "Grafting from" Method ,Macromolecules, 38, pp. 7330-7336.
12. Na, K., and Bae, Y.H., (2005), Polymeric Drug Delivery Systems, Taylor & Francis, Boca Raton, pp. 129-194.
13. Lee, E.S., Na, K. and Bae, Y.H., (2003), Polymeric micelle for tumor pH and folate-mediated targeting, Journal of Controlled Release, 91(1-2), pp. 103-113.
14. Kim, B., La Flamme, K. and Peppas, N.A., (2003), Dynamic Swelling Behavior of pH-Sensitive Anionic Hydrogels Used for Protein Delivery, J. Appl. Polym. Sci., 89, pp. 1606-1613.
15. Khayat, Z., Griffiths, P.C., Grillo, I., Heenan, R.K., King, S.M. and Duncan, R., (2006), Characterising the size and shape of polyamidoamines in solution as a function of pH using neutron scattering and pulsed-gradient spin-echo NMR, International Journal of Pharmaceutics, 317(2), pp. 175-186.
16. Ringsdorf, H., Venzmer J. and Winnik F.M. (1991), Interaction of hydrophobically-modified poly-N-isopropylacrylamides with model membranes - or playing a molecular accordion, Angewandte Chemie (International Edition in English), 30 (3), pp. 315-318.
17. Ramkissoon-Ganorkar, C., Liu, F., Baudys, M. and Kim, S.W., (1999), Effect of molecular weight and polydispersity on kinetics of dissolution and release from pH/temperature-sensitive polymers, J. BIOM. SC. P., 10(10), pp. 1149-1161.
18. Rodríguez-Cabello, J.C., Reguera, J., Girotti, A., Arias, F.J. and Alonso, M., (2006), Genetic Engineering of Protein-Based Polymers: The

Example of Elastinlike Polymers, Advances in Polymer Science, 200, pp 119-167.

19. Dreher, M.R., Raucher, D., Balu, N., Colvin, O., Ludeman, S.M. and Chilkoti A., (2003), Evaluation of an elastin-like polypeptide-doxorubicin conjugate for cancer therapy, J. Control Release, 91(1-2), pp. 31-43.
20. Furgeson, D.Y., Dreher, M.R. and Chilkoti, A., (2006), Structural optimization of a "smart" doxorubicin-polypeptide conjugate for thermally targeted delivery to solid tumors, Journal of Controlled Release, 110(2), pp. 362-369.
21. Matsumura, Y. and Maeda, H., (1986), A New Concept for Macromolecular Therapeutics in Cancer Chemotherapy: Mechanism of Tumoritropic Accumulation of Proteins and the Antitumor Agent Smancs, Cancer Res., 46, pp. 6387-6392.
22. Richardson, S.C.W., Pattrick, N.G., Stella, Y. K., Ferruti, P. and Duncan, R., (2001), Poly(Amidoamine)s as Potential Nonviral Vectors:? Ability to Form Interpolyelectrolyte Complexes and to Mediate Transfection *in vitro*, Biomacromolecules, 2(3), pp 1023-1028.
23. Qiu L.Y. and Bae Y.H., (2006), Polymer architecture and drug delivery, Pharm. Res., 23(1), pp 1-30.
24. Brocchini S., Duncan R., (1999), Encyclopaedia of Controlled Drug Delivery, Mathiowitz E., editor. Wiley; New York, NY. pp. 786.
25. Duncan, R. and Spreafico, F., (1994), Polymer conjugates. Pharmacokinetic considerations for design and development, Clin. Pharmacokinet., 27(4), pp. 290-306.
26. Seymour, L.W., Duncan, R., Strohalm, J. and Kopecek, J., (1987), Effect of molecular weight (Mw) of N-(2-hydroxypropyl) methacrylamide copolymers on body distribution and rate of excretion after subcutaneous, intraperitoneal, and intravenous administration to rats, J. Biomed. Mater. Res., 21(11), pp. 1341-1358.
27. Duncan R., (2003), The dawning era of polymer therapeutics, Nat. Rev. Drug Discov., 2(5), 347-360.
28. Suortti, T., Gorenstein, M.V. and Roger, P., (1998), Determination of the molecular mass of amylose. Journal of Chromatography A, 828, 515-521.
29. Donald, A. M., (1994), Physics of foodstuffs, Reports on Progress in Physics, 57, pp. 1081-1135.
30. Marques, A.P., Rui L. Reis. and John A. Hunt, (2005), An In Vivo Study of the Host Response to Starch-Based Polymers and Composites Subcutaneously implanted in rats, Macromol. Biosci., 5 , pp. 775-785.
31. Kozlowska, M, Fryder, K. and Wolko, B., (2001), Peroxidase involvement in the defense response of red raspberry to

Didymellaapplanata (Niessl/Sacc.), ActaPhysiologiaePlantarum, 23(3), pp 303-310.

32. Martins, A., Chung, S., Pedro, A. J, Sousa, R. A., Marques, A. P., Reis, R. L. and Neves, N. M., (2009), Hierarchical starch-based fibrous scaffold for bone tissue engineering applications, Journal of Tissue Engineering and Regenerative Medicine, 3(1), pp. 37-42.
33. Azevedo, H. S. and Reis, R. L., (2009), Encapsulation of a-amylase into starch-based biomaterials: An enzymatic approach to tailor their degradation rate, Acta Biomaterialia, 5(8), pp. 3021-3030.
34. Reddy, N. and Yang, Y., (2009), Preparation and properties of starch acetate fibers for potential tissue engineering applications, Biotechnology and Bioengineering ,103(5), pp. 1016-1022.
35. Tuzlakoglu, K., Pashkuleva, I., Rodrigues, M. T., Gomes, M. E., van Lenthe, G. H., Muller, R. and Reis, R. L., (2010), A new route to produce starch-based fiber mesh scaffolds by wet spinning and subsequent surface modification as a way to improve cell attachment and proliferation, Journal of Biomedical Materials Research Part A, 92A(1), pp. 369-377.
36. Alves, N. M., Leonor, I. B., Azevedo, H. S., Reis, R. L. and Mano, J. F., (2010), Designing biomaterials based on biominerali-zation of bone, J. Mater. Chem., 20, pp. 2911-2921.
37. Torres, F. G., Troncoso, O. P., Grande, C. G. and Díaz, D. A., (2011), Biocompatibilty of starch-based films from starch of Andean crops for biomedical applications, Materials Science and Engineering: C, 31(8), pp. 1737-1740.
38. BeMiller, J. N., (2009), Starch: Chemistry and Technology (Food Science and Technology), 3rdedition, pp. 17-18.
39. Waliszewski, K. N., Aparicio, M. A., Bello, L. A. and Monroy, J. A., (2003), Changes of banana starch by chemical and physical modification, Carbohydrate Polymers, 52(3), pp. 237-242.
40. Gustavson B., (1983), Cadexomer Iodine: Introduction. In Cadexomer Iodine. Fox J.A., Fisher H. (eds), Stuttgart: Schattauer Verlag, pp. 35-41.
41. Sangseethong, K., Lertphanich, S. and Sriroth, K., (2009), Physico-chemical Properties of Oxidized Cassava Starch Prepared under Various Alkalinity Levels, Starch – Stärke 61(2), pp. 92-100.
42. Lawal, O.S., Adebowale, K.O., Ogunsanwo, B. M., Barba, L.L. and Ilo, N.S., (2005), Oxidized and acid thinned starch derivatives of hybrid maize: functional characteristics, wide-angle X-ray diffractometry and thermal properties, International Journal of Biological Macromolecules, 35(1-2), pp. 71-79.
43. Kesler, C.C. and Hjermstad, E.T., Preparation of starch ethers in original granule form. US Pat. 2, 516, 633.

44. Shefer, A., Shefer, S., Kost, J. and Langer, R., (1992), Structural characterization of starch networks in the solid state by cross-polarization magic-angle-spinning carbon-13 NMR spectroscopy and wide angle x-ray diffraction, Macromolecules, 25 (25), pp 6756-6760.
45. Van Soest, J. J. G. and Vliegenthart, J. F.G, (1997), Crystallinity in starch plastics: consequences for material properties, Trend in biotechnology, 15(6), pp. 208-213.
46. Malafaya P.B., Stappers F. and Reis R.L., (2006), Starch-based microspheres produced by emulsion crosslinking with a potential media dependent responsive behavior to be used as drug delivery carriers, J. Mater. Sci. Mater. Med., 17(4), pp.371-377.
47. Kost, J. and Shefer, S., (1990), Chemically-modified poly-saccharides for enzymatically-controlled oral drug delivery, 11(9), pp.695-698.
48. Shalviri, A., Liu, Q., Abdekhodaie, M. J. and Wu, X. Y., (2010), Novel modified starch–xanthan gum hydrogels for controlled drug delivery: Synthesis and characterization, Carbohydrate Polymers, 79(4), pp. 898-907.
49. Reis, A.V., Guilherme, M. R., Moia, T. A., Mattoso, L. H. C., Muniz, E. C. and Tambourgi, E. B., (2008), Synthesis and chara-cterization of a starch-modified hydrogel as potential carrier for drug delivery system, Journal of Polymer Science Part A: Polymer Chemistry, 46(7), pp. 2567-2574.
50. Simi C. K. and Abraham, E., (2007), Hydrophobic grafted and cross-linked starch nanoparticles for drug delivery, Bioprocess Biosyst. Eng., 30(3), pp. 173-80.
51. Shaikh, M. M. and Lonikar, S. V., (2009), Starch–acrylics graft copolymers and blends: Synthesis, characterization, and applications as matrix for drug delivery, Journal of Applied Polymer Science 114(5), pp. 2893-2900.
52. Yu, L., Dean, K. and Li, L., (2006), Polymer blends and composites from renewable resources, Progress in Polymer Science, 31(6), June 2006, pp. 576-602.
53. Breslin, V. T. and Li, B., (2003), Weathering of starch–polyethylene composite films in the marine environment, Journal of Applied Polymer Science, 48(12), pp. 2063-2079.
54. Raghavan, D. and Emekalam, A., (2001), Characterization of starch/ polyethylene and starch/polyethylene/poly(lactic acid) composites, Polymer Degradation and Stability,72(3), pp. 509-517.
55. St-Pierre, N., Favis, B.D., Ramsay, B.A., Ramsay, J.A. and Verhoogt, H., (1997), Processing and characterization of thermoplastic starch/ polyethylene blends, Polymer, 38(3), pp. 647-655.

56. Danjaji, I.D., Nawang, R., Ishiaku, U.S., Ismail, H. and Z.A.M MohdIshak, (2002), Degradation studies and moisture uptake of sago-starch-filled linear low-density polyethylene composites, Polymer Testing, 21(1), pp. 75-81.
57. Warth, H., Mülhaupt, R. and Schätzle, J., (1997), Thermoplastic cellulose acetate and cellulose acetate compounds prepared by reactive processing, Journal of Applied Polymer Science, 64(2), pp. 231-242.
58. Arvanitoyannis, I. and Biliaderis, C. G., (1999), Physical properties of polyol-plasticized edible blends made of methyl cellulose and soluble starch, Carbohydrate Polymers, 38(1), pp. 47-58.
59. Pereira, C.S., Cunha, A.M., Reis, R.L., Vázquez B. and San Román J, (1998), New starch-based thermoplastic hydrogels for use as bone cements or drug-delivery carriers, J. Mater. Sci. Mater. Med., 9(12), pp. 825-33.
60. Kozlowska, A., Kozlowski, M. and Iwanczuk, A. (2005) Starch-based multiphase polymer materials. e-Polymerse-Polymers. Volume 5, Issue 1, Pages 577-584.
61. Bajpai A. K. and Shrivastava J., (2007), Studies on alpha-amylase induced degradation of binary polymeric blends of crosslinked starch and pectin, J. Mater. Sci. Mater. Med., 18(5), pp. 765-777.
62. Soares, G.A., De Castro, A.D., Cury, B.S. and Evangelista, R.C., (2013), Blends of cross-linked high amylose starch/pectin loaded with diclofenac, Carbohydr. Polym., 91(1), pp.135-42.
63. Silva, I., Gurruchaga, M. and Goñi, I., (2009), Physical blends of starch graft copolymers as matrices for colon targeting drug delivery systems, Carbohydrate Polymers, 76(4), pp. 593-601.
64. Torres, F. G., Arroyo, O. H., Grande, C. and Esparza, E., (2006), Bio- and Photo-degradation of Natural Fiber Reinforced Starch-based Biocomposites, International Journal of Polymeric Materials, 55(12), pp. 1115-1132.
65. Gómez, C., Torres, F. G., Nakamatsu, J. and Arroyo, O. H., (2006), Thermal and Structural Analysis of Natural Fiber Reinforced Starch-Based Biocomposites, International Journal of Polymeric Materials, 55(11), pp. 893-907.
66. Marques, A.P., Reis, R.L. and Hunt, J.A., (2002), The biocompatibility of novel starch-based polymers and composites: in vitro studies, Biomaterials, 23(6), pp. 1471–1478.
67. Saikia, J.P., Banerjee, S., Konwar, B.K. and Kumar, A., (2010), Biocompatible novel starch/polyaniline composites: characteri-zation, anti-cytotoxicity and antioxidant activity. Colloids Surf. B: Biointerfaces., 81(1), pp.158-64.

68. Torres, F. G., Grande, C. J., Gomez, C. M., Troncoso, O. P., Canet-Ferrer, J. and Martínez-Pastor, J., (2009), Development of self-assembled bacterial cellulose–starch nanocomposites, Materials Science and Engineering: C, 29(4), pp. 1098-1104.
69. Chen, B. and Evans, J. R.G., (2005), Thermoplastic starch–clay nanocomposites and their characteristics, Carbohydrate Polymers, 61(4), pp. 455-463.
70. Lee, P. I., (1985), Kinetics of drug release from hydrogel matrices, Journal of Controlled Release, 2, pp. 277-288.
71. Yoon, H.-S, Lee, J.H., S.-T. and Li, H.,(2009), Utilization of retrograded waxy maize starch gels as tablet matrix for controlled release of theophylline, Carbohydrate Polymers, 76, pp. 449-453.
72. Li, Y., de Vries, R., Slaghek, T., Timmermans, J., Cohen, M.A. and Norde, W., (2009), Preparation and Characterization of Oxidized Starch Polymer Microgels for Encapsulation and Controlled Release of Functional Ingredients, Biomacromolecules, (7), pp 1931-1938.
73. Li, Y., Zhang, Z., van Leeuwen, H.P., Cohen, M.A., Norde, W. and Kleijn, J.M., (2011a), Uptake and release kinetics of lysozyme in and from an oxidized starch polymer microgel, Soft Matter, 7, pp. 10377-10385.
74. Li, Y., Kleijn, J.M., Cohen, M.A,. Slaghek, T., Timmermans, J. and Norde, W., (2011b), Mobility of lysozyme inside oxidized starch polymer microgels, Soft Matter, 7, pp. 1926-1935.
75. Ali, A.E-H. and AlArifi, A., (2009), Characterization and in vitro evaluation of starch based hydrogels as carriers for colon specific drug delivery systems, 78(4), pp. 725-730.
76. Bajpai, S. K. and Saxena, S., (2004), Enzymatically degradable and pH-sensitive hydrogels for colon-targeted oral drug delivery. I. Synthesis and characterization, Journal of Applied Polymer Science, 92(6), pp. 3630-3643.
77. Elvira, C., Mano, J.F., San Roman, J and Reis, R.L.,(2002), Starch-based biodegradable hydrogels with potential biomedical applications as drug delivery systems, Biomaterials, 23 (9) pp. 1955-1966.
78. Frisch, H. L., (1980), Sorption and Transport in Glassy Polymers – A Review, Polymer Engineering & Science, 20(1), pp. 2-13.
79. Crank J., (1975), The Mathematics of Diffusion, Clarendon Press, Oxford.
80. Sadeghi, M. and Hosseinzadeh, H., (2008), Synthesis of Starch – Poly(Sodium Acrylate-co-Acrylamide) Superabsorbent Hydrogel with Salt and pH-Responsiveness Properties as a Drug Delivery System, Journal of Bioactive and Compatible Polymers , 23(4), pp. 381-404.

81. Shaheen, S. M., Takezoe, K. and Yamaura, K. (2004), Effect of binder additives on terbutaline hydrogels of a-PVA/NaCl/H2O system in drug delivery: I. Effect of gelatin and soluble starch, Bio-Medical Materials and Engineering, 14(4), pp. 371-382.

82. Husseini, G.A. and Pitt, W.G., (2008), Micelles and Nanoparticles for Ultrasonic Drug and Gene Delivery, Adv. Drug Delivery, 60(10), pp. 1137-1152.

83. Besheer, A., Hause,G., Kressler,J. and Mäder, K., (2007), Hydrophobically Modified Hydroxyethyl Starch: Synthesis, Characterization, and Aqueous Self-Assembly into Nano-Sized Polymeric Micelles and Vesicles, Biomacromolecules, 8 (2), pp 359-367.

84. Zhang, A., Zhang, Z., Shi, F., Ding, J., Xiao, C., Zhuang, X., He, C., Chen, L. and Chen, X., (2013), Disulfide crosslinked PEGylated starch micelles as efficient intracellular drug delivery platforms, Soft Matter, 9, pp. 2224-2233.

85. Ju, B., Yan, D. and Zhang, S., (2012), Micelles self-assembled from thermoresponsive 2-hydroxy-3-butoxypropyl starches for drug delivery, Carbohydrate Polymers, 87(2), pp. 1404-1409.

86. Chang, T.M.S., (1964), Semipermeable Microcapsules, Science, 146, pp. 524-525

87. Lim, F. and Sun, A.M., (1980), Microencapsulated islets as bioartificial endocrine pancreas, Science, 210, pp. 908-910.

88. Langer R. (1998), Drug delivery and targeting, Nature., 392 (6679), pp. 5-10.

89. Park, K., (1997), Controlled Drug Delivery: Challenges and Strategies, American Chemical Society, Washintong DC, pp. 629.

90. Li, B.-Z., Wang, L.-J., Li, D., Adhikari, B. and Mao, Z.H.,(2012), Preparation and characterization of crosslinked starch microspheres using a two-stage water-in-water emulsion method, Carbohydrate Polymers, 88(3), pp. 912-916.

91. Elfstrand, S., Lagerlöf, J., Hedlund, K. and Mårtensson, A., (2008), Carbon routes from decomposing plant residues and living roots into soil food webs assessed with 13C labelling, Soil Biology and Biochemistry, 40(10), pp. 2530-2539.

92. Fang, Y.-Y., Wang, L.-J., Li, D., Li, B.-Z., Bhandari, B., Chen, X.D. and Mao, Z.-H., (2008), Preparation of crosslinked starch microspheres and their drug loading and releasing properties, Carbohydrate Polymers, 74(3), pp. 379-384.

93. Mundargi, R.C., Shelke, N.B., Rokhade, A.P., Patil, S.A. and Aminabhavi T.M., (2008), Formulation and in-vitro evaluation of novel starch-based

tableted microspheres for controlled release of ampicillin, Carbohydrate Polymers, 71(1), pp. 42-53.

94. Fundueanu, G., Constantin, M., Ascenzi,P. and Simionescu, B.C., (2010), An intelligent multicompartmental system based on thermo-sensitive starch microspheres for temperature-controlled release of drugs, Biomedical Microdevices, 12(4), pp 693-704.
95. Baimark, Y., Srisa-ard, M. and Srihanam, P., (2010), Morphology and thermal stability of silk fibroin/starch blended microparticles, eXPRESS Polymer Letters, 4(12), pp. 781-789.
96. Tuovinen, L., Peltonen, S., Liikola, M., Hotakainen, M., Lahtela-Kakkonen, M., Poso, A. and Järvinen, K., (2004), Drug release from starch-acetate microparticles and films with and without incorporated a-amylase, Biomaterials, 25(18), pp. 4355-4362.
97. Balmayor, E.R., Tuzlakoglu, K., Azevedo, H.S. and Reis, R.L., (2009), Preparation and characterization of starch-poly-e-caprolactone microparticles incorporating bioactive agents for drug delivery and tissue engineering applications, Acta Biomaterialia, 5(4), pp. 1035-1045.
98. Desai, K.G., (2006), Properties of Tableted High-Amylose Corn Starch–Pectin Blend Microparticles Intended for Controlled Delivery of Diclofenac Sodium, J. Biomater. Appl., 21(3), pp. 217-233.
99. Rodrigues, A., and Emeje, M., (2012), Recent applications of starch derivatives in nanodrug delivery, Carbohydrate Polymers, 87(2), pp. 987-994.
100. Panyam, J., and Labhasetwar, V., (2003), Biodegradable nanoparticles for drug and gene delivery to cells and tissue, Advanced Drug Delivery Reviews, 55(3), pp. 329-347.
101. Hughes, G.A., (2005), Nanostructure-mediated drug delivery, Nanomedicine: Nanotechnology, Biology and Medicine, 1(1), pp. 22-30.
102. Guzmán, K., Taylor, M.R. and Banfield, J.F., (2006), Environ-mental Risks of Nanotechnology: National Nanotechnology Initiative Funding, 2000-2004, Environ. Sci. Technol., 40 (5), pp 1401-1407.
103. Shalviri, A., Chan, H.K., Raval, G., Abdekhodaie, M.J., Liu, Q., Heerklotz, H. and Wu, X.Y., (2013), Design of pH-responsive nanoparticles of terpolymer of poly(methacrylic acid), polysorbate 80 and starch for delivery of doxorubicin, Colloids and Surfaces B: Biointerfaces, 101, pp. 405-413.
104. Shi, A.M., Li, D., Wang, L.J. and Adhikari, B., (2012), Rheological properties of suspensions containing cross-linked starch nanoparticles prepared by spray and vacuum freeze drying methods, Carbohydrate Polymers, 90(4), pp. 1732-1738.

105. Dandekar, P., Jain, R., Stauner, T., Loretz, B., Koch, M., Wenz, G. and Lehr, C.M., (2012), A hydrophobic starch polymer for nanoparticle-mediated delivery of docetaxel., Macromol Biosci., 12(2), pp. 184-94.
106. Arockianathan, P.M., Sekar, S., Sankar, S., Kumaran, B. and Sastry, T.P., (2012), Evaluation of biocomposite films containing alginate and sago starch impregnated with silver nano particles, Carbohydrate Polymers, 90(1), pp. 717-724.
107. Karrout, Y., Neut, C., Wils, D., Siepmann, F., Deremaux, L., Flament, M. P., Dubreuil, L., Desreumaux, P. and Siepmann, J., (2011), Peas starch-based film coatings for site-specific drug delivery to the colon, Journal of Applied Polymer Science, 119(2), pp. 1176-1184.
108. Wang, Q., Hu, X.W., Du, Y.M. and Kennedy, J.F., (2010), Alginate/starch blend fibers and their properties for drug controlled reléase, Carbohydrate Polymers, 82(3), pp. 842-847.

3 Seaweed Polysaccharides in Advanced Drug Delivery Application to Carragennans

S. Lefnaoui* and N. Moulai-Mostefa

Faculty of Sciences and Technology, University of Medea, Ain D'Heb, 26001 Medea, Algeria.

3.1 Introduction

The active ingredients are rarely administered directly but in the form of preparations or formulated drugs. The pharmaceutical forms contain, in addition to the active pharmaceutical ingredient, a number of components in order to facilitate the manufacturing process and drug delivery.

Formulations that are able to control the release of drug have become an integral part of the pharmaceutical industry and in particular the oral drug delivery systems. These formulations were investigated in the last years due to their many benefits over conventional dosage. For the formulation of such systems, excipients are currently included in the determination of new forms for specific functions and, in some cases; they directly or indirectly influence the rate of drug release.

Pharmaceutical excipients are substances that have been evaluated for their safety and are carefully included in a release of active ingredient. The vehicle is designed to facilitate drug administration and for stability, bioavailability and efficacy[1].

Polymers have been successfully used in the formulation of different dosage forms and are specifically useful in the design of modified release drug delivery systems. Both synthetic and natural polymers were investigated extensively for this purpose[2,3], but the use of natural polymers for pharmaceutical applications is attractive because they are economical, non-toxic, biodegradable and biocompatible[4,5].

Traditionally, excipients were included in drug formulations as inert vehicles that provided the necessary weight, consistency and volume for the correct administration of the active ingredient, but in modern

* ***Corresponding Author***

pharmaceutical dosage forms they often fulfill multi-functional roles such as improvement of the stability, release and bioavailability of the active ingredient, and performance of technological functions that ensure ease of manufacture.

The specific application of natural polymers in pharmaceutical formulations include their use in the manufacture of solid monolithic matrices, implants, films, beads, microparticles, nanoparticles, inhalable and injectable systems as well as viscous liquid formulations[6,7]. Within these dosage forms, polymeric materials have fulfilled different roles such as binders, thickeners, stabilizers, emulsifiers, suspending agents, gelling agents and bioadhesives[2].

Polymers are often utilized in the design of novel drug delivery systems such as those that target delivery of the drug to a specific region in the gastrointestinal tract or in response to external stimuli to release the drug. This can be done via different mechanisms including coating of tablets with polymers having pH dependent solubilities or incorporating non-digestible polymers that are degraded by bacterial enzymes in the colon[8].

Polysaccharides as natural polymers are the most abundant renewable resource on the earth. They are designed by nature to carry out various specific functions. Generally, polysaccharides are highly functional polymers with magnificent structural diversity and functional versatility.

Polysaccharides are a class of biopolymers constituted by simple sugar monomers. The monomers (monosaccharides) are linked together by O-glycosidic bonds that can be made to any of the hydroxyl groups of a monosaccharide, conferring polysaccharides the ability to form both linear and branched polymers. Differences in the monosaccharide composition, chain shapes and molecular weight dictate their physical properties including solubility, gelation and surface properties. These biological polymers can be obtained from different sources, microbial, animal and vegetal[9].

3.2 Seaweed polysaccharides

The study of marine biodiversity, particularly polysaccharides, represents a major challenge for research. It is called "blue chemistry", by analogy with "green chemistry". So it is important to explore the marine biodiversity, especially marine polysaccharides because they are very promising for the future, especially in therapeutics.

There are two types of sources of polysaccharides of marine origin: algae and bacteria (Fig. 3.1). The use of algal polysaccharides is widely deployed in industry (food, cosmetics, biotechnology, pharmaceutical and agriculture)[10].

These polymers from three different categories (Table 3.1) of macro-algae (green, red, brown) are generally exploited for their textural properties (gelling agent, thickener, stabilizer, etc.). They also exhibit anticoagulant and/or antiviral activities. Include for example carrageenans and agars (extracted from red algae), mainly used for their texturing properties. Other algal polysaccharides with similar properties are common, such as alginates (brown algae) or ulvans (green algae).

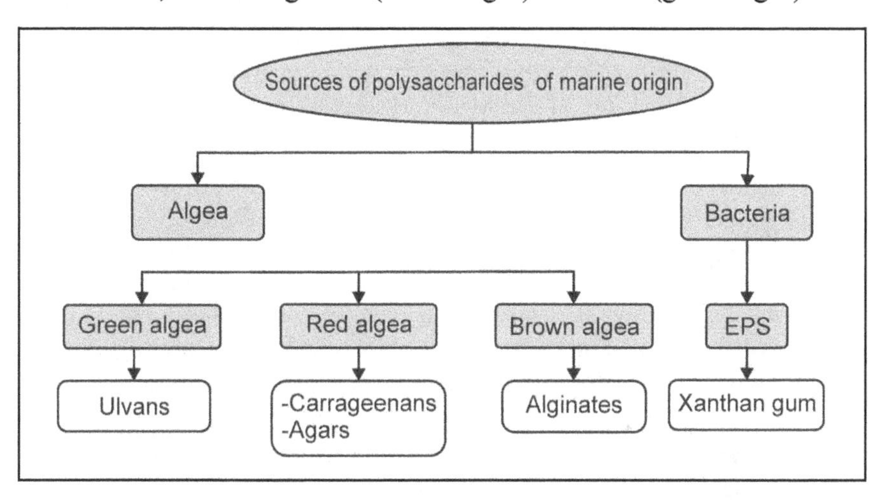

FIGURE 3.1 Types of sources of polysaccharides of marine origin.

In addition to these important sources of polysaccharides, some bacteria are capable of producing exopolysacharides (EPS), such as *Vibriodiabolicus* and *Alteromonasinfernus*. The best biopolymer is EPS xanthan used in the food and pharmaceutical industry. Other important polysachharides as chitin and chitosan are derived from the exoskeleton of marine crustaceans. These seaweed polysaccharides polymers have been widely proposed as scaffold materials in tissue engineering applications as well as carriers for drug delivery systems as described in more detail in the following sections.

Seaweeds are rich sources of sulfated polysaccharides, including some ones that have become valuable additives in the food industry because of their rheological properties as gelling and thickening agents (e.g., carrageenan). In addition, sulfated polysaccharides are recognized to possess a number of biological activities including anticoagulant, antiviral, and immuno-inflammatory activities that might find relevance in functional food, cosmetic and pharmaceutical applications.

The complex polysaccharides from the brown, red and green seaweeds possess broad spectrum therapeutic properties. Considering the immense biomedical prospects of sulfated polysaccharides, the profound and emerging functional properties published in recent times will be discussed here.

Several algal species have been recognized as crucial sources of sulfated polysaccharides (SP). These SP constitute an important ingredient of cell walls and get harvested by suitable extraction or precipitation method, followed by purification, characterization and biological studies. Currently, the regenerative medicine and tissue engineering application of sulfated polysaccharides has become a hot research area.

Carrageenans are a family of linear SP, extracted from red seaweeds, viz. Gracialaria, Gigartina, Gelidium, Lomentaria, Corallina, Champia, Solieria, Gyrodinium, Nemalion, Sphaerococcus, Boergeseniella, Sebdenia and Scinaia[11]. Three categories of carrageenans have been identified based on their sulfation degree, solubility and gelling properties[12]. Ulvan is the major water soluble, sulfated polysaccharide, extracted from the cell wall of green algae, i.e., Ulva, Enteromorpha, Monostroma, Caulerpa, Codium, Gayralia. Ulvans are composed of disaccharide repetition moieties made up of sulfated rhamnose linked to glucuronic acid, iduronic acid, or xylose[13].

3.2.1 Alginates

Alginate is a well known polysaccharide obtained from natural sources, such as its extraction from cell walls and intercellular spaces of marine brown algae, and its production by bacteria. It can be characterized as an anionic copolymer where its chemical structure (Fig. 3.2) is based on a backbone of (1-4) linked β- D-mannuronic acid (M units) and α-L-guluronic acid (G units) of widely varying composition and sequence depending on the source of alginate. Alginate has a variable molecular weight, depending on the enzymatic control during its production and the degree of depolymerization caused by its extraction. Typically, commercial alginates have an average molecular weight of approximately 200,000 Da, but alginates with values as high as 400,000-500,000 Da are also available[14].

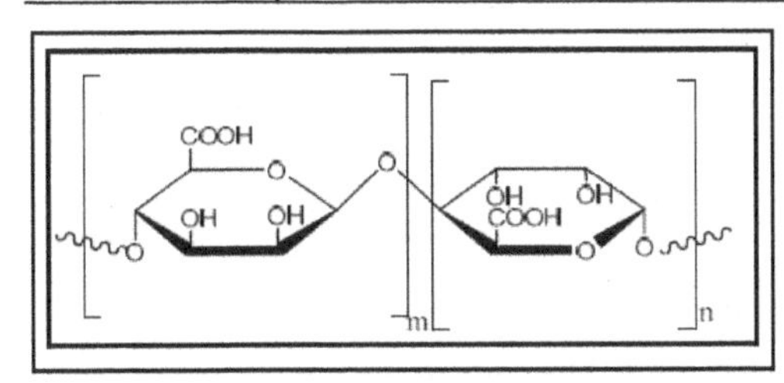

FIGURE 3.2 Chemical structure of alginate.

The amount of homopolymeric and heteropolymeric blocks present in alginate depends upon the sources previously mentioned. Thermo-irreversible gels can be formed by controlling the amount of calcium and acidity. Although temperature does not hinder gelation of sodium alginate, it affects the final gel properties[15].

Alginate is the most common encapsulating agent used for probiotic bacteria, which is mainly attributed to its ease of formation of capsules and release of probiotic bacteria using chelation in phosphate buffer. Alginate capsules are made using different techniques like extrusion and emulsification.

Alginate with high content of guluronic acid block can produce, in the form of calcium salts, cross-links stabilizing the structure of the polymer in a rigid gel form. This property enables alginate solutions to be processed into the form of films, beads and sponges.

Alginate has a large number of free hydroxyl and carboxyl groups distributed along the backbone, which are highly reactive and turn it into an ideal candidate for being appropriately modified by chemical functionalization. These chemical modifications have been achieved using techniques such as oxidation, sulfation, esterification, amidation, grafting methods[16].

The biocompatibility behavior and the high functionality make alginate a favorable biopolymer material for its use in biomedical applications, such as scaffolds in tissue engineering, immobilization of cells and controlled drug release devices[17,18].

Alginates are used for various applications in drug delivery, such as in matrix type alginate gel beads, in liposomes, in modulating gastrointestinal transit time, for local applications and to deliver the biomolecules in tissue engineering applications[19].

Bioadhesive sodium alginate microspheres for intranasal systemic delivery were prepared, as an alternative therapy to injection, and to obtain improved therapeutic efficacy in the treatment of hypertension and angina pectoris[20].

In case of its applications in nanomedicine, alginate has also been extensively investigated as a drug delivery device in which the rate of drug release can be modified by varying the drug polymer interaction, as well as by chemical immobilization of the drug in the polymer backbone using the reactive carboxylate groups[21].

Alginates have been used for immobilization of Langerhans islets in the treatment of experimental diabetes mellitus in rats[22] and for microencapsulation of hormone-producing cell for the treatment of diabetes mellitus and parathyroid disease[23]. It was shown that alginates led to monocytes and macrophages stimulation, resulting in an increase of cytokine production. The macrophage stimulation seems to be related to alginates ability to enhance the healing process. Experimental data suggested that low-molecular weight alginates could be useful in the prevention of obesity, hypercholesterolemia, diabetes and antitumor[24,25].

3.2.2 Carageenans

Carrageenans are sulphated marine hydrocolloids obtained by extraction from seaweeds of the class Rhodophyceae, represented by *Chondruscrispus, Euchemaspinosum, Gigartinaskottsbergi, Gigartina-stellata, Iradaealaminariodes*. Carrageenan is not assimilated by the human body. It provides only bulk but no nutrition. It has been categorized into three types known as kappa (κ), iota (ι) and lambda (λ). It was observed that λ-carrageenan produces viscous solutions but does not form gels, while κ-carrageenan forms a brittle gel and ι-carrageenan produces elastic gels[26]. Their chemical structures are shown in Fig. 3.3.

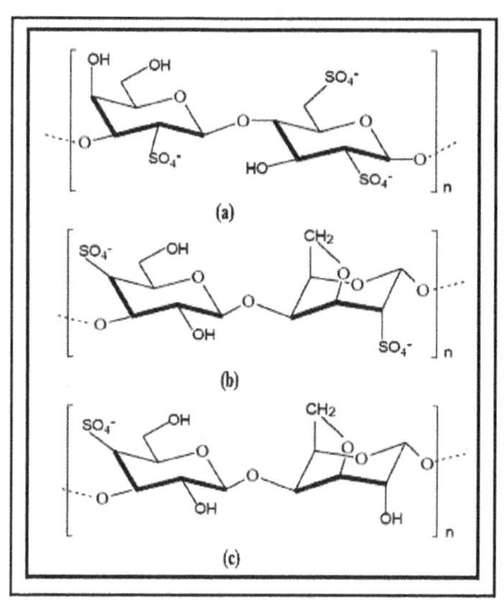

FIGURE 3.3 Chemical structures of (a) λ-carrageenan, (b) ì-carrageenan and (c) κ-carrageenan.

The three types of carrageenan do not exist singly, but as a combination of two types and available with one predominating type or molecules containing structural components of more than one type. Each type of carrageenan has a unique set of characteristics, including gel strength, viscosity, temperature stability, synergism, and solubility. The solubility of each carrageenan depends on the number of sulfate groups, which increases water solubility, compared with anhydro bridges, which is hydrophobic. Having the least water solubility, κ-carrageenan has one sulfate group for every two galactose units and one anhydro bridge; ι-carrageenan has two sulfate groups for every two galactose units along with one anhydro bridge; and with the highest solubility, λ-carrageenan has three sulfate groups for every two galactose units and no anhydro bridges[27]. There are distinct differences among each type of carrageenan, but all types are soluble at high temperatures[28].

Carrageenans are suitable in the formulation of controlled release tablets[28,29]. It was shown that matrices made of ι-carrageenan and λ-carrageenan sustained the release of three different model drugs and showed release profiles that approached zero order kinetics. It was found that factors such as tablet diameter, drug to carrageenan ratio and ionic strength of the dissolution medium may play a role in the release of drug from these matrices[30].

Hydrogel beads prepared from a mixture of cross-linked κ-carrageenan with potassium and cross-linked alginate were used as novel carriers for controlled drug delivery systems[31].

3.2.3 Agar Gum

Agar gum is extracted from the red-purple marine algae of the Rhodophyceae class. The species include *Gelidiumcartilagineum* and *Gracilariaconfervoides*[27]. Its chemical structure is shown in Fig. 3.4.

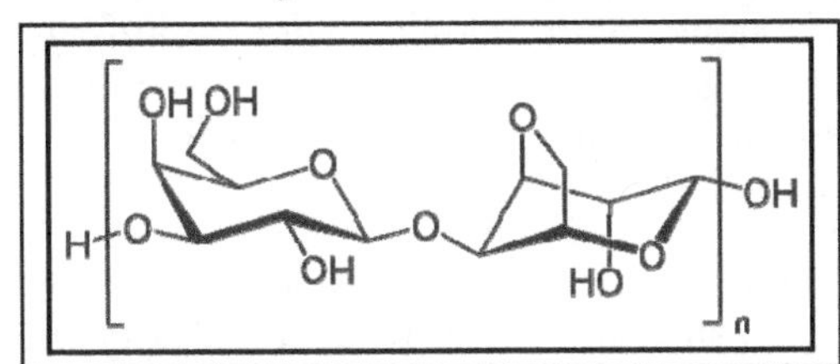

FIGURE 3.4 Chemical structures of an agarose polymer.

Agar (or agar-agar) is a phycocolloid, which is constructed from complex saccharide molecules[32]. Agar and its variant agarose contain also variable amounts of sulfate, piruvate and uronate substituents. Agar is insoluble in cold water but is soluble in boiling water. Agar dissolved in hot water and permitted to cool will form thermally reversible gels without the need of acidic conditions or oxidizing agents. This characteristic gives agars the ability to perform a reversible gelling process without losing their mechanical and thermal properties. The significant thermal hysteresis of the gel is another important property for commercial applications. The gelling process in agar is due to the formation of hydrogen bonds in a continuous way[33]. Gelation occurs as a result of a coil-double helix transition[34]; helices interact among themselves and the gel is formed by linked bundles of associated right-handed double helices. The resulting three-dimensional network is capable of immobilizing water molecules in its interstices.

Agar has pharmaceutical industrial applications as suspending agent for radiological solutions (barium sulfate), as a bulk laxative as it gives a smooth and non-irritating hydrated bulk in the digestive tract, and as a formative ingredient for tablets and capsules to carry and release drugs. Pharmaceutical grade agar has a viscous consistency. In microbiology, agar is the medium of choice for culturing bacteria on solid substrate.

The possibility to use agar and agarose beads for sustained release of water soluble drugs has been investigated[35]. Agarose has a significantly lower sulfate content, better optical clarity and increased gel strength with respect to agar, but it is considerably more expensive[36]. Agar beads instantaneously form by gelification[37]. The results of dissolution and release studies indicated that agar beads could be useful for the preparation of sustained release dosage forms, although no many further studies have been developed.

3.2.4 Ulvans

The green alga is the most diverse group of algae and, it is a paraphyletic group. Because they are aquatic and manufacture their own food, these organisms are called "algae," along with certain members of the Chromista, the Rhodophyta, and photosynthetic bacteria, even though they do not share a close relationship with any of these groups[38].

Ulvan is the major water-soluble polysaccharide found in green seaweed. It contains sulfate, rhamnose, xylose, iduronic and glucuronic acids as main constituents[38]. Ulvan structure shows great complexity and variability as evidenced by the numerous oligosaccharides repeating structural units identified in native and chemically modified ulvan preparations[39]. The main repeating disaccharide units reported are ulvanobiouronic acid 3-sulfate types containing either glucuronic or iduronic acid[40] as shown in Fig. 3.5.

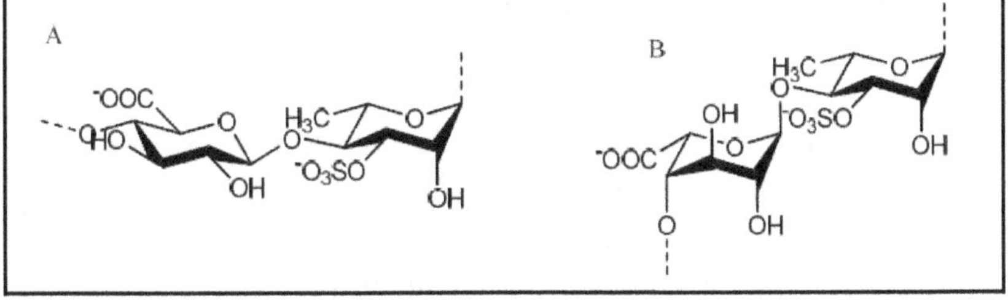

FIGURE 3.5 Chemical structure of ulvan.

The medicinal interest in green algae is centred in its polysaccharidic part, particularly ulvan, to be used as a therapeutic active agent. The presence of glucuronic and iduronic acids makes ulvan a very special polysaccharide. This fact gains importance as this biopolymer is used in pharmaceutical and biomedical applications. Another remarkable property of ulvan is its sulphation degree. Sulphate groups have long been associated with different biological activities[41,42]. The use of ulvan as a strategic alternative to various synthetic or animal bioactive agents would take advantage of its algal origin, together with high availability and low expected production costs, low cytotoxicity and broad spectrum of biological activities. It could be applied as an antiviral agent, antioxidant, as an anti-coagulant alternative to heparin, anti-hyperlipidemic or of its anti-proliferative activity towards cancer cells or for therapy for diseases where the immune system is impaired[43,44]. Furthermore, due to its similarity with mammalian glycosaminoglycans, it could be exploited as a pharmaceutical where the delivery of glycosaminoglycans is needed, such as for the treatment of musculoskeletal disorders[45]. On the other hand, rhamnose moieties ubiquitous in the ulvan backbone, as mentioned above, may be the basis for its use for the treatment of skin pathologies, particularly the ones related with age and its effects[45].

As ulvan is readily recognized by hepatocyte membrane receptors, it could be used as a biomaterial for diagnostic or therapeutic purposes[46]. Moreover, and taking advantage of the ability of ulvan to complex with metal ions, it can find applications where the removal of these ions from the body is required[47].

The technological development of ulvan is still in the field of possibilities and mainly focused on its applicability as a biomaterial for tissue engineering and regenerative medicine. Within the context of tissue engineering and regenerative medicine, the main objective is to guide cells into forming a functional living tissue[48].

3.2.5 Microbial Polysaccharides

Bacterial polysaccharides has potential applications in the pharmaceutical industry. Although many known marine bacteria produce exopolysaccharides (EPS). Natural polysaccharide gums have also been obtained as carbohydrate fermentation products including xanthan gum, produced in pure culture fermentation by the bacteria *xanthomona-scampestris*[49].

Xanthan gum is a complex microbial exopolysaccharide produced from glucose fermentation by *xanthomonas campestris pv.*, a plant bacterium. The gum consists of D-glucosyl, D-mannosyl, and D-glucuronyl acid residues. It also contains *o*-acetyl and pyruvyl residues in variable proportions[50]. Xanthan gum is an acidic polysaccharide gum of penta-saccharide subunits (Fig. 3.6).

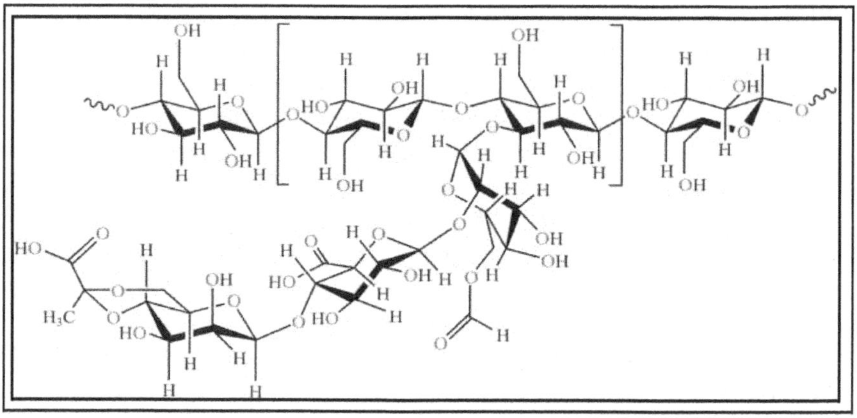

FIGURE 3.6 Chemical structures of Xanthan gum.

Xanthan gum is non-toxic and has been approved by the Food and Drug Administration (FDA) for its use as food additive without quantity limitations[51]. Xanthan has been used in a wide range of industries including food, cosmetics and pharmaceutical industries.

3.2.6 Chitosan

Chitosan is the deactelyated form of chitin, the most abundant natural biopolymer after cellulose. Chitin is the major structural component of exoskeleton of invertebrates and the cell walls of fungi[52]. Chitosan is a linear polysaccharide composed by units of glucosamine and N-acetylglucosamine linked by (1→4) β-glycosidic bonds. It is a hydrophilic biopolymer obtained industrially by hydrolysing the aminoacetyl groups of chitin (the main component of the shells of crab, shrimp and krill) by an alkaline deacetylation treatment[53]. The chemical structures of chitin and chitosan are shown in Fig. 3.7.

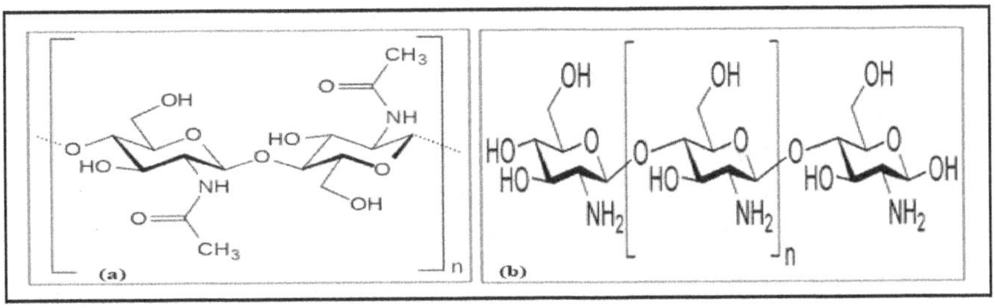

FIGURE 3.7 Chemical structure of (a) chitin, (b) chitosan.

Chitosan is a biocompatible polymer; it was used in many applications such as food industry, environment and pharmaceutical[54]. It has been assayed as biomaterial for wound healing and prosthetic material, since it can be biodegraded by enzyme action[55]. As a cationic polymer, it was used as a nasal drug delivery. Because of its bioadhesive properties, chitosan has received substantial attention as carrier in novel bioadhesive drug delivery systems which prolong the residence time of drugs at the site of absorption and increase the drug bioavailability[56]. Thus, some drugs administered via nasal or gastrointestinal routes have improved their treatment efficacy when they are included into chitosan-based systems[57]. Chitosan appears to be a promising matrix for the controlled release.

Despite its biocompatibility, the applications of chitosan are limited due to its insolubility above pH 6. Chitosan is a weak base and it is insoluble in water and organic solvents. However, it is soluble in diluted aqueous acidic solution. It is possible to increase the solubility of chitosan in water removing one or two hydrogen atoms from the amino groups of chitosan, and introducing some hydrophilic segments[58].

3.3 Polymeric Drug Delivery Systems

The conventional oral and intravenous routes of drug administration do not provide ideal pharmacokinetic profiles especially for drugs, which display high toxicity and/or narrow therapeutic windows. For such drugs the ideal pharmacokinetic profile will be one where in the drug concentration reached therapeutic levels without exceeding the maximum tolerable dose and maintains these concentrations for extended periods. One of the methods to deliver drugs in an ideal case would be by encapsulating the drug in a polymer matrix. The advantages that polymeric drug delivery products can offer are localized delivery of drug, sustained delivery of drugs and their stabilization.

Drug–polymer systems may also be useful in protecting the drug from biological degradation prior to its release. The development of these devices starts with the use of non-biodegradable polymers, which rely on the diffusion process, and subsequently progresses to the use of biodegradable polymers, in which swelling and erosion take place.

Based on the physical or chemical characteristics of polymer, drug release mechanism from a polymer matrix can be categorized in accordance to three main processes[59]:

- Drug diffusion from the non-degraded polymer (diffusion-controlled system).
- Enhanced drug diffusion due to polymer swelling (swelling-controlled system).
- Drug release due to polymer degradation and erosion (erosion-controlled system).

In all three systems, diffusion is always involved. For biodegradable polymer matrix, release is normally controlled by the hydrolytic cleavage of polymer chains that lead to matrix erosion, even though diffusion may be still dominant when the erosion is slow. This categorization allows mathematical models to be developed in different ways for each system.

Mathematical modeling of drug release provides insights concerning mass transport and chemical processes involved in drug delivery system as well as the effect of design parameters, such as the device geometry and drug loading, on drug release mechanism.

The release of drug is affected by multiple factors such as the physicochemical properties of the solutes, the structural characteristics of the material system, release environment, and interactions between these factors (Fig. 3.8).

In general, solute diffusion, polymeric matrix swelling, and material degradation are suggested to be the main driving forces for solute transport from drug containing polymeric matrices[60]. Fick's law of diffusion provides the description of solute transport from polymeric matrices where the Fickian diffusion refers to the solute transport process[61]. The purpose of mathematical modeling is to simplify the complex release process and to gain insight into the release mechanisms of a specific material system. Thus, a mathematical model mainly focuses on one or two dominant driving forces. Models have been developed to describe solute transport based on each of the different mechanisms. However, disconnects exist between theories and experimental data since there are multiple driving forces involved in a single transport process. Moreover, the existing mathematical models may be insufficient in describing more complex material systems.

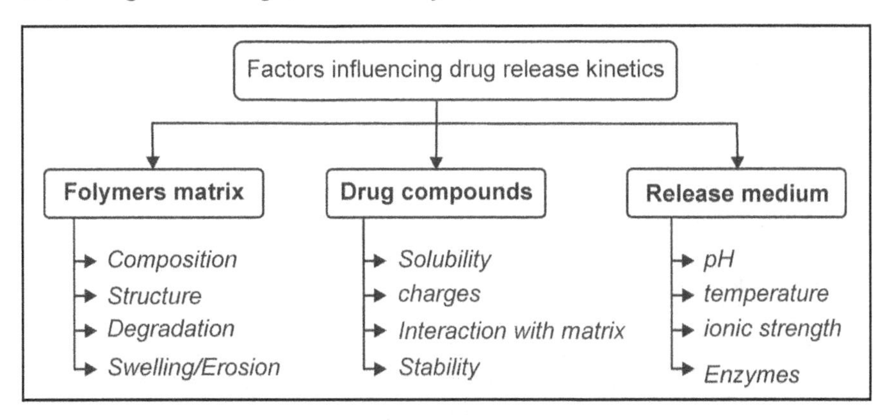

FIGURE 3.8 Factors influencing drug release in polymeric matrix.

There are several types of drug release systems that are designed according to the mechanism they employ. These include controlled dissolution, diffusion, combination of dissolution and diffusion, ion exchange and transport control.

The investigation on the phenomena involved in the swelling and drug release process for systems containing polymeric materials is important for the design of delivery systems. It has been shown that the polymeric content, which is related to swelling behaviour, and the viscosity grade are the determining factors in predicting the drug release from hydrophilic matrices[62]. The purpose of orally administered hydrophilic matrices is generally to prolong drug delivery with zero-order kinetics in order to maintain a constant *in vivo* plasma concentration and constant pharmacological effect[63]. The mechanisms by which drugs released are complex and involve different processes:

- entry of the aqueous medium into the matrix,
- matrix swelling,
- drug dissolution in the medium,
- drug diffusion through the gel layer,
- matrix erosion.

Mathematical models are used in order to describe the drug release kinetics. These models provide a basis for the study of mass transport mechanisms that are involved in the drug release. In general, diffusion, erosion, and degradation are the most important mechanisms for solute transport from polymeric matrices.

Factors that could affect drug release through matrices may be due to several sources such as characteristics of drug loading, polymers, formulation factors, presence of other drugs, presence of other excipients and manufacturing process.

Drugs with a low molecular weight tend to diffuse through the gel layer more easily than those of high molecular weight. Water-soluble drugs tend to follow a release mechanism based on diffusion through the gel layer, while water-insoluble drugs do so mainly through a mechanism of erosion[64]. Thus, water insoluble drugs tend to have slower release rates. Model-independent and model-dependent kinetic analyses have revealed the existence of a behavioral trend in the release mechanism of drugs from hydrophilic matrices as a function of the drug solubility[65]. Another factor is the drug particle size. This factor affects the diffusion of the compound through the hydrated layer of the polymer and hence the release rate. In the case of highly water-soluble drugs, drug particle size affected the rate of release because the actual drug form channels that cross the gel layer with the consequent increase of the system porosity. The higher the percentage of drug incorporated into the matrix, the greater the porosity.

The good characteristics of polysaccharides, considered as excellent vehicles for modulating drug release rates, are a result of their structure and composition. The characteristics that determine whether polysaccharides will behave in one way or another are mainly their molecular weight, the structure of the chain (linear or branched) and the side chains of each chain monomer[66]. In addition, the structural and physicochemical characteristics of the polysaccharide are decisive in the drug release mechanism, some will be more suitable than others, depending on the aim pursued and the drug desired. The molecular weight of the polymer is directly related to gel strength and is of great importance in drug release since it is decisive for the passage of water through the gel layer during swelling[65]. For hydrophilic matrices, gel strength is what determines the erosion capacity of the polymer, such that the higher the molecular weight of the polymer, the greater the degree of swelling and the lower its ability to erode. This characteristic, together with the solubility of the drug incorporated into the polymer matrix, will govern the release mechanism.

Normally, high-molecular weight polymers increase the viscosity of the solution. This means that viscosity is one of the parameters that controls drug release and its mechanism. Some authors have reported that an increase in the percentage of polymer in the matrix corresponds to a decrease in the drug release rate[66]. These observations are consistent with the findings of Mitchell et al.[67], who reported that an increase in the percentage of polymer elicits a greater degree of cross-linking of the polymer side chains, which in turn decreases the viscosity and increases the concentration of the gel and its tortuosity, which are the main characteristics that prevent the diffusion of the drug through the gel. However, some authors do not agree with this notion, as is the case, of Tiwari et al.[68].

3.4 Potential use of κ-carrageenan in Pharmaceutical Applications

3.4.1 Introduction

Kappa-carrageenan (κ-carr) is mostly used in the food industry as gelling, stabilizing agent and thickener because of its high hydrophilicity, mechanical strength, biocompatibility and biodegradability. Carrageenans are widely utilized due to their excellent physical functional properties. They are also used in various non-food products, such as pharmaceutical, cosmetics, printing and textile formulations[69].

After the inclusion of carrageenan in the US Pharmacopoeia with its own monograph, it was used in many investigations in pharmaceutical technology[70,71].

In particular κ-carrageenan was investigated in gels[72,73]; films[74]; tablets[75]; capsules[76]; and pellets[77]. Further studies are carried out on the effects of ions and their interactions[78].

They have proved to be useful as tableting excipients due to the good compatibility, high robustness and persistent viscoelasticity of the tablet during compression. Kappa-carrageenan would reduce or eliminate toxicity in biomedical applications. For this reason, it has been applied for immobilizing protein and controlled drug delivery systems[79]. These interesting properties indicated that carrageenans are suitable excipients for sustained-release formulations[80].

Carrageenans are also used for their biological activities as agents for the induction of experimental inflammation and inflammatory pains[81]. They have also shown several potential pharmaceutical properties including antitumor, immunomodulatory[82], and anticoagulant activities[83]. Some *in vitro* studies suggest that carrageenans may also have antiviral properties, inhibiting the replication of herpes and hepatitis A viruses[84].

3.4.2 κ-carrageenan as a Gelling Agent

Hydrogels are polymeric materials that have a three dimensional network structure and can swell considerably in aqueous medium without dissolution. They can be prepared in a wide-variety of physical forms[85].

Recent advances in hydrogel technology have focused on finding more biocompatible, non-toxic material intended for pharmaceutical, biomedical or even in food application. Hydrogels formed from polysaccharides, such as carrageenan (κ-Carr), are good candidates for drug release systems to their nontoxicity, their easy gelling ability, thermo reversibility of the gel network and viscoelastic properties[86].

In order to increase the stability of hydrogel carrier for drug delivery, the polymeric material needs to be cross-linked. Several cross-linking reagents have been used for cross-linking such as glutaraldehyde, tripolyphosphate, ethylene glycol, and genipin.

Some studies have shown that cross-linked hydrogels exhibit different properties as compared to uncross-linked hydrogels.

Because of the ionic nature of the polymer, gelation of carrageenans is strongly influenced by the presence of electrolytes. κ- Carrageenan forms a gel with potassium ions, but also shows gelation under salt-free conditions. However, gels prepared in the presence of metallic ions were substantially stronger than those obtained under salt-free conditions[87].

The viscosity of carrageenans depends on concentration, temperature, presence of other solutes, type of carrageenan and molecular weight.

An experimental study made on the development and evaluation of swelling and adhesion of kappa carrageenan-based gels was carried out where the k-Carr concentrations were varied between 0.5 and 0.7%, so as to form gels without syneresis phenomenon (k-Carr concentration > 0.7%), and more consistent than colloidal solutions (concentration in k-Carr < 0.5%). It was shown that the degree of gel swelling increases with increasing polymer concentration. However, before reaching a concentration of 0.3%, the solutions are too diluted and do not form a gel. Beyond 0.3%, the solutions are more semi-dilute, and then concentrated to 0.5%, forming a gel, the absorbing solvent which surrounds and swells as a function of time.

κ-Carr is present in an orderly propellers which have the property to aggregate and form a network responsible for the gel form. Gelling temperatures and melting is about 23°C and 34°C, respectively, regardless of the concentration of κ-carrageenan, the total ion concentration of the medium is kept constant. From 0.6% gels become too concentrated, and the appearance of the phenomenon of syneresis can be observed, the swells less.

The polymers containing hydroxyl, carboxyl or sulfate like κ-Carr seem to be the most favorable bioadhesion candidates with mucus glycoprotein network (Fig. 3.9). This phenomena can be explained either by an electron transfer during the contact κ-carr - mucus, which leads in turn to the formation of a double layer of electric charges at the interface, and then to the theory of adsorption, which calls for a mechanism for participation by secondary chemical bonds (Van der Waals forces, hydrogen bonds).

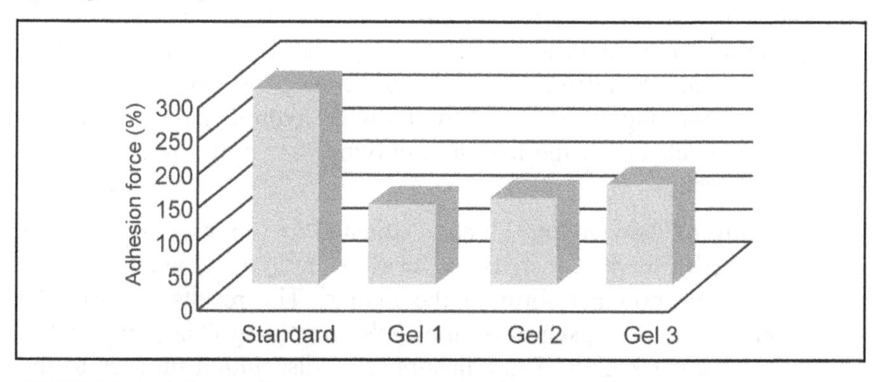

FIGURE 3.9 Adhesion force of detachment of k-carrageenan-based gels (Gel 1: 0.5%, Gel 2 0.55%, Gel 3: 0.6%).

3.4.3 Carrageenans as a Blend Agent

Hydrogels made of single polymers have been extensively used in drug delivery studies. However, in many cases a single polymer alone cannot meet divergent demands in controlled drug release in terms of both properties and performance[88].

Polymer blending is an attractive method to obtain combined advantages of different polymers such as physical and mechanical properties. The resultant hydrogels exhibit different properties than those of the original polymers and the polymer blends may show synergistic properties[89,90].

Tablets made from mixtures of κ-carrageenan and sodium carboxymethylcellulose and κ-carrageenan and hydroxypropylmethylcellulose have been shown to release drug at a constant rate over a period of time[91].

In this chapter, we present some results on the evaluation of combination of two polysaccharides. The objective is to improve bioadhesion and bioavailability of an antifungal pregelatinized starch (PGS)-based gel containing miconazole by adding a second polysaccharide (κ-carrageenan) as a polymer blend. The study of the effect of κ-carrageenan (κ-Carr) on the hydrophilic matrix based-PGS was performed by evaluating bioadhesion, swelling kinetics and drug release kinetics in different dissolution media.

The results showed that the swelling of hydrogels is governed by two phenomena: the diffusion of the external environment for hydration gel and relaxation of the polymer chains. The swelling capacity of gels containing κ-Carr is significantly better than that of gels based on PGS alone. This can be attributed to the difference in molecular weight between the two polymers caused by the wide hydrodynamic volume of any molecular chains of the hydrated polymer having a high molecular weight (κ-Carr).

The study of the swelling kinetics adjusted to Davidson and Peppas model showed a marked increase in swelling ratio as a function of the variation of the polymer forming the matrix. The results confirm the surface erosion with a faster erosion of gels containing PGS compared to the gel containing κ-Carr. The adjustment of dissolution profiles to the power law model shows an anomalous transport mechanism ($n > 0.5$) for gels containing κ-Carr, characteristic of a combination of Fickian diffusion and relaxation of the polymer chains.

The adhesion strength varies between 0.22 N for the PGS-based gel and achieved 0.56N for the gel containing 0.7% κ-Carr (Fig. 10). The increase in adhesion strength gels containing increasing concentration of κ-Carr may be related to the formation of a stronger gel detachment and penetrating deeper in contact with the membrane[92].

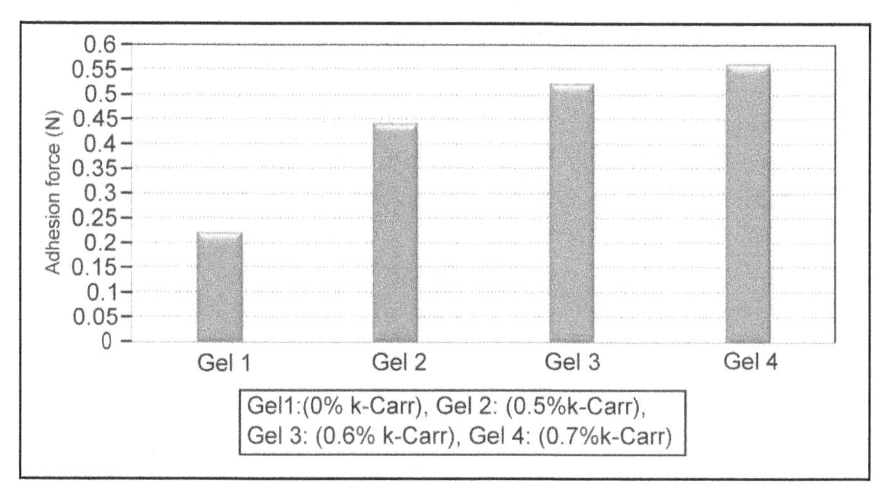

FIGURE 3.10 Effects of introduction of k-Carr on the bioadhesive properties of a PGS-based gel.

The influence of polymer concentration in improving of adhesion strength is related to the formation of stronger ties with the membrane, which can be explained by the length and the number of polymer chains penetrating into the membrane layer. When the polymer concentration is too low, the number of polymer chains penetrating per unit volume of mucus is low, and the interaction between the polymer and the mucus is unstable. In general, a more concentrated polymer would result in a longer length and penetrating chain offering better adhesion. However, for each polymer, there is a critical concentration beyond which the polymer produces a state of equilibrium with the substrate. Consequently, the accessibility of the solvent to the polymer network, and the penetration of the polymer chain is substantially reduced. Therefore, higher concentrations of polymers do not necessarily improve, and in some cases reduces the bioadhesive properties.

From the *in vitro* dissolution tests, it was shown that the effect of addition of κ-Carr to PGS-based gel is very important. Indeed the combination of polymers (PGS: κ-Carr) allowed us to obtain a slow and gradual release of the active ingredient.

The addition of κ-Carr as a blend in the formulation increases the bond strength of the gel, as well as improves its absorption and prolongs the time to saturation and therefore delays erosion of the polymer matrix formed by the combination of PGS and κ-Carr.

Gels formed by the two polymers showed a higher rate of swelling, this confirms the synergy between the two polymers and reinforcing polymer network formation of entangled chains between PGS and κ-Carr. The drug release through the hydrophilic polymer matrix depends not only on its ability to spread through the hydrogel, but also by the drug dissolution and the polymer swelling.

3.4.4 Modified Kappa Carrageenan as an Excipients Agent

Many studies have been conducted to modify the physical and chemical properties of carrageenan[93,94]. Chemical modification of polysaccharides is considered as one of the most important route to enhance the properties of these biopolymers. Recent trends and strategies in research are geared towards functionalization of known materials and carboxymethylation of polysaccharides is one of the widely studied conversions that lead to development of new biomaterials with very promising applications. Carboxymethylation reactions have been studied in natural polymers, such as cellulose, chitin, chitosan, dextran, and xylanas starting materials. The derivates obtained are polyelectrolytes that can be applied in the chemical, pharmaceutical, food and cosmetic industries[95].

To extend the use of κ-carrageenan, the industrial and scientific interests in the carboxymethylation of κ-carrageenan have increased significantly in recent years. Recently, novel materials based on carboxymethylated kappa-carrageenan were successfully developed as suitable carriers for intestinal-targeted delivery of bioactive molecules[96,97].

The degree of substitution of polysaccharides can be increased by a multi-step reaction, i.e., the carboxymethylated polysaccharide synthesized is isolated and subsequently carboxymethylated again under comparable conditions. Multi-step carboxymethylation was applied as an alternative path of polysaccharide etherification leading to new functionalization patterns and derivatives with valuable features[98].

Synthesis of carboxymethyl κ-carrageenan can avoid the problems of the synthesis of sulfated polysaccharides, and can also provide great advantages such as technical simplicity, low cost and environmental protection. Moreover, after carboxymethylation, the carboxymethyl κ-carrageenan would contain sulfate and carboxyl groups, as the nearest structural analogues of heparin.

Conclusion

Although excipients have traditionally been included in formulations as inert substances to mainly make up volume and assist in the manufacturing process, they are included in dosage forms to fulfill specialized functions for improved drug delivery. Majority of investigations on natural polymers in drug delivery systems center concern polysaccharides.

As the polysaccharides excipients are promising biodegradable materials, these can be chemically compatible with the excipients in drug delivery systems. In addition polysaccharides excipients are non-toxic, freely available, capable of chemical modifications, and less expensive compared to their synthetic counterparts. They have a major role to play in pharmaceutical industry. Therefore, in the future, there is going to be continued interest in the natural excipients to have better materials for drug delivery systems.

Polysaccharides can also be modified to have tailor-made products for drug delivery systems. They have found application not only in sustaining of drug release but are also proving useful for development of gastro retentive dosage forms, bioadhesive systems and microcapsules.

References

1. The Joint IPEC-PQG Good Manufacturing Practices Guide for Pharmaceutical Excipients, 2006.
2. Guo, J., Skinner, G.W., Harcum, W.W., Barnum, P.E. 1998. Pharmaceutical applications of naturally occurring water-soluble polymers. PSTT 1, 254-261.
3. Varshosaz, J., Tavakoli, N., Eram, S.A. 2006. Use of natural gums and cellulose derivatives in production of sustained release Metoprolol tablets. Drug Deliv. 13, 113-119.
4. Satturwar, P.M., Fulzele, S.V., Dorle, A.K. 2003. Biodegradation and *in vivo* biocompatibility of rosin: a natural film-forming polymer. AAPS PharmSciTech. 4, 1-6.
5. Chivate, A.A., Poddar, S.S., Abdul, S., Savant, G. 2008. Evaluation of Sterculia foetida gum as controlled release excipient. AAPS PharmSciTech. 9, 197-204.
6. Chamarthy, S.P., Pinal, R. 2008. Plasticizer concentration and the performance of a diffusion-controlled polymeric drug delivery system. Colloids Surf. A. Physiochem. Eng. Asp 331, 25-30.

7. Alonso-Sande, M., Teijeiro, D., Remuñán-López, C., Alonso, M.J. 2008. Glucomannan, a promising polysaccharide for biopharmaceutical purposes. Eur. J. Pharm. Biopharm. 72, 453-462.
8. Xing, L., Dawei, C., Liping, X., Rongqing, Z. 2003. Oral colon-specific drug delivery for bee venom peptide: development of a coated calcium alginate gel beads-entrapped liposome. J. Control. Rel. 93, 293-300.
9. Cascone, M.G., Barbani, N., Cristallini, C., Giusti, P., Ciardelli, G., Lazzeri, L. 2001. Bioartificial polymeric materials based on polysaccharides. J. Biomater. Sci. 12, 267-281.
10. Venugopal, J., Ramakrishna, S. 2005. Applications of polymer nanofibers in biomedicine and biotechnology. Appl. Biochem. Biotechnol. 125, 147-157.
11. Tuvikene, R., Truus, K., Vaher, M., Kailas, T., Martin, G., Kersen, P. 2006. Extraction and quantification of hybrid carrageenans from the biomass of the red algae Furcellaria lumbricalis and Coccotylus truncatus. Proc. Estonian Acad. Sci. Chem. 55, 40-53.
12. Leibbrandt, A., Meier, C., Ko¨nig-Schuster, M., Weinmu¨llner, R., Kalthoff, D., Pflugfelder, B., Graf, P., Frank-Gehrke, B., Beer, M., Fazekas, T., Unger, H., Prieschl-Grassauer, E., Grassauer, A. 2010. Iotacarrageenan is a potent inhibitor of influenza A virus infection. PLoS One 5(12):e14320. doi:10.1371/journal.pone.0014320.
13. Lahaye, M., Robic, A. 2007. Structure and functional properties of ulvan, a polysaccharide from green seaweeds. Biomacromolecules. 8, 1765-1774.
14. Rehm, B.H.A. (Ed). (2009). Alginates: Biology and Applications, Springer, 978-3-540-92678.
15. Onsoyen, E. 1997. Thickening and Gelling Agents for Food. 2nd ed. Imeson A, ed. New York, Chapman & Hall, 22-44.
16. Yang, J. S., Xie, Y. J., He, W. 2011. Research progress on chemical modification of alginate: A review. Carbohydr. Polym. 84, 33-39.
17. Barbosa, M., Granja, P., Barrias, C., Amaral, I. 2005. Polysaccharides as scaffolds for bone regeneration. ITBM-RBM 26, 212-217.
18. Pandey, R., Ahmad, Z. 2011. Nanomedicine and experimental tuberculosis: facts, flaws, and future. Nanomedicine 7, 259-272.
19. Rajinikanth, PS., Sankar, C., Mishra, B. 2003. Sodium alginate microspheres of metoprolol tartrate for intranasal systemic delivery: Development and evaluation. Drug Deliv. 10, 21-28.

20. Fuchs-Koelwel, B., Koelwel, C., Gopferich, A., Gabler, B., Wiegrebe, E. 2004. Tolerance of a new calcium-alginate-insert for controlled medication therapy of the eye. Ophthalmologe 101, 496-499.
21. Nair, L. S., Laurencin, C. T. 2007. Biodegradable polymers as biomaterial. Prog. Polym. Sci. 6, 762-798.
22. Fan, M.Y., Lum, Z.P., Fu, X.W., Levesque, L., Tai, I.T., Sun, A.M. 1990. Reversal of diabetes in BB rats in transplantation of encapsulation pancreatic islets Diabetes, 39, 519-522.
23. Soon-Shiong, P.R.E., Henitz, N., Merideth, Q.X., Yao, T., Zheng, M., Murphy. 1994. Insulin independence in a type I diabetic patient after encapsulated islet transplantation Lancet. Rev. Diabet. Stud. 343, 950-951.
24. Kimura,Y., Watanabe, K., Okuda, H. 1996. Effects of soluble sodium alginate on cholesterol excretion and glucose tolerance in rats. J. Ethnopharmacol. 54, 47-54.
25. Fujiihara, M., Nagumo, T. 1993. An influence of the structure of alginate on the chemotactic activity of macrophage and the antitumor activity. Carbohydr. Res. 243, 211-216.
26. Sudhakar, Y., Kuotsu, K., Bandyopadhyay, AK. 2006. Buccal bioadhesive drug delivery – A promising option for orally less efficient drugs. J. Control Rel. 114, 15-40.
27. Imeson, A. P. 2000. Handbook of Hydrocolloids. Phillips G O,Williams P A, eds. Boca Raton; CRC Press LLC, 87-102.
28. Picker, K.M. 1999. Matrix tablets of carrageenans. I. A compaction study. Drug Dev. Ind. Pharm. 25, 329-337.
29. Mohamadnia, Z., Zohuriaan-Mehr, M.J., Kabiri, K., Jamshidi, A., Mobedi, H. 2008. Ionically crosslinked carrageenan-alginate hydrogel beads. J Biomater. Sci. Polym. 19, 47-59.
30. Gupta, V.K., Hariharan, M., Wheatley, T.A., Price, J.C. 2001. Controlled-release tablets from carrageenans: effect of formulation, storage and dissolution factors. Eur. J. Pharm. Biopharm. 51, 241-248.
31. Bonferoni, M.C., Rossi, S., Tamayo, M., Pedrez, J.L., Dominguez, G., Caramella, C. 1993. On the employment of λ-carrageenan in a matrix system. I. Sensitivity to dissolution medium and comparison with Na carboxymethylcellulose and xanthan gum. J Control. Rel. 26, 119-127.
32. Araki, C. 1966. Some recent studies on the polysaccharides of agarophytes. Proc. Int. Seaweed Symp. 5, 3-19.

33. Stephen, A.M., Phillips, G.O., Williams, P.A. 1995. Food Polysaccharides and Their Applications; Marcel Dekker: New York, NY, USA, 187-203.
34. Morris, V.J. 1986. Gelation of polysaccharide. In Functional Properties of Food Macromolecules, Mitchell, J.A., Ledwards, D.A., Eds.; Elsevier: London, UK, 121-170.
35. Nakano, M., Nakamur, Y., Takikawa, K., Kouketsu, M., Arita, T. 1979. Sustained release of sulfamethizole from agar beads. J. Pharm. Pharmacol. 31, 869-872.
36. Kojima, T., Hashida, M., Muranishi, S., Sezaki, H. 1978. Antitumor activity of timed-release derivative of mitomycin C, agarose bead conjugate. Chem. Pharm. Bull. 26, 1818-1824.
37. El-Raheem El-Helw, A., El-Said, Y. 1988. Preparation and characterization of agar beads containing phenobarbitone sodium. J. Microencapsul. 5, 159-163.
38. Percival, E., McDowell, R.H. 1967. Chemistry and Enzymology of Marine Algal Polysaccharides; Academic Press, New York, NY, USA, 219.
39. Lahaye, M., Brunel, M., Bonnin, E. 1997. Fine chemical structure analysis of oligosaccharides produced by an ulvan-lyase degradation of the water-soluble cell-wall polysaccharides from Ulva sp. (Ulvales, Chlorophyta). Carbohydr. Res. 304, 325-333.
40. Guangling, J., Guangli, Y., Junzeng, Z.H., Stephen, E. 2011. Chemical Structures and Bioactivities of Sulfated Polysaccharides from Marine Algae. Mar. Drugs 9, 196-223.
41. Pengzhan, Y., Ning, L., Xiguang, L., Gefei, Z., Quanbin, Z., Pengcheng, L. 2003a. Antihyperlipidemic effects of different molecular weight sulfated polysaccharides from Ulva pertusa (Chlorophyta). Pharmacol. Res. 48, 543-549.
42. Leiro, J.M., Castro, R., Arranz, J.A., Lamas, J. 2007. Immunomodulating activities of acidic sulphated polysaccharides obtained from the seaweed Ulva rigida C. Agardh. Int. Immunopharmacol. 7, 879-888.
43. Wijesekara, I., Pangestuti, R., Se-Kwon Kim, S-K. 2011. Biological activities and potential health benefits of sulfated polysaccharides derived from marine algae. Carbohydr. Polym. 84, 14-21
44. El-Baky, H.H.A., Baz, F.K.E., Baroty, G.S.E. 2009. Potential biological properties of sulphated polysaccharides extracted from the macroalgae Ulva lactuca L. Acad. J. Cancer Res., 2, 01-11.

45. Faury, G., Molinari, J., Rusova, E., Mariko, B., Raveaud, S., Huber, P., Velebny, V., Robert, A.M., Robert, L. 2011. Receptors and aging: structural selectivity of the rhamnose-receptor on fibroblasts as shown by Ca^{2+} -mobilization and gene-expression profiles. Arch. Gerontol. Geriat 53, 106-112.

46. Massarelli, I., Murgia, L., Bianucci, A.M., Chiellini, F., Chiellini, E. 2007. Understanding the selectivity mechanism of the human asialoglycoprotein receptor (ASGP-R) toward gal- and man-type ligands for predicting interactions with exogenous sugars. Int. J. Mol. Sci. 8, 13-28.

47. Lahaye, M., Robic, A. 2007. Structure and functional properties of ulvan, a polysaccharide from green seaweeds. Biomacromolecules 8, 1765-1774.

48. Mano, J.F., Silva, G.A., Azevedo, H.S., Malafaya, P.B., Sousa, R.A., Silva, S.S., Boesel L.F., Oliveira, J.M., Santos, T.C., Marques, A.P., Neves, N.M., Reis, R.L. 2007. Natural origin biodegradable systems in tissue engineering and regenerative medicine: present status and some moving trends. J. Roy. Soc. Interf. 4, 999-1030.

49. Parija, S., Misra, M., Mohanty, A.K. 2001. Studies of natural gum adhesive extracts - An overview. Polym. Rev. 4, 175-197.

50. Rosalam, S., England, R. 2006. Review of xanthan gum production from unmodified starches by Xanthomonas comprestris sp. Enzyme Microbiol. Technol. 39, 197-207.

51. Garcia-Ochoa, F., Santos, V.E., Casas, J.A., Gomez, E. 2000. Xanthan gum: production, recovery, and properties. Biotechnology Advances, 18, 549-579.

52. Shahidi, F., Kamil, J., Arachichi, V., Jeon, Y.J. 1999. Food applications of chitin and chitosans. Trends Food Sci. Technol. 10, 37-51.

53. Muzzarelli, R.A.A., Muzzarelli, C. 2005. Chitosan chemistry: Relevance to the biomedical sciences Polysaccharides 1: Structure, characterization and use. Adv. Polym. Sci. 186, 151-209.

54. Ghaouth, E.A., Arul, J., Asselin, A., Benhamou, N. 1992. Antifungal activity of chitosan on post harvest pathogens: induction of morphological and cytological alterations and rhizopus stolonifer. Mycol. Res. 96, 769-779.

55. Soto-Perlata, N.V., Muller, H., Knorr, D. 1989. Effect of chitosan treatments on the clarity and color of apple juice. J. Food Sci. 54, 495-496.

56. Varum, F.J., McConnell, E.L., Sousa, J.J., Veiga, F., Basit, A. W. 2008. Mucoadhesion and the gastrointestinal tract. Crit. Rev. Ther. Drug Carrier Syst. 25, 207-258.
57. Guerrero, S., Teijón, C., Muñiz, E., Teijón, J. M., Blanco, M. D. 2010. Characterization and *in vivo* evaluation of ketotifen-loaded chitosan microspheres. Carbohydr. Polym. 79, 1006-1013.
58. Srinophakun, T., Boonmee, J. 2011. Preliminary Study of Conformation and Drug Release Mechanism of Doxorubicin-Conjugated Glycol Chitosan, via cis-Aconityl Linkage, by Molecular Modeling. Int. J Mol. Sci. 12, 1672-1683.
59. Narasimhan, B., Peppas, N.A. 1997. Molecular analysis of drug delivery systems controlled by dissolution of the polymer carrier, J. Pharm. Sci. 86, 297-304.
60. Artifin, D.Y., Lee, L.Y., Wang, C.H. 2006. Mathematical modeling and simulation of drug release from microspheres: implication to drug delivery systems. Adv. Drug. Deliv. Rev. 58, 1274-1325.
61. Grassi, M., Grassi, G. 2005. Mathematical modeling and controlled drug delivery: matrix systems. Curr. Drug Deliv. 2, 97-116.
62. Ranga-Rao, K.V., Padmalatha, D.K., Buri, P. 1990. Influence of molecular size and water solubility of the solute on its release from swelling and erosion controlled polymeric matrices. J. Control Rel. 12, 133-141.
63. Liu, B.T., Hsu, J-P. 2005. Inward release polymer matrix covered by a permeable membrane: a possible zero-order controlled release device. Chem. Eng. Sci. 60, 5803-5808.
64. Bettini, R., Catellani, P.L., Santi, P., Massimo, G., Peppas, N.A., Colombo, P. 2001. Translocation of drug particles in HPMC matrix gel layer: effect of drug solubility and influence on release rate. J. Control Rel. 70, 383-391.
65. Tahara, K., Yamamoto, K., Nishihata, T. 1996. Application of model-independent and model analysis for the investigation of effect of drug solubility on its release rate from hydroxypropyl methylcellulose sustained release tablets. Int. J. Pharm. 133, 17-27.
66. Brady, J.E., Dürig, T., Shang, S.S., in: Qiu, Y., Liu, L., Chen, Y., Zhang, G.G.Z., Porter, W. 2009. Developing Solid Oral Dosage Forms Pharmaceutical Theory & Practice, Esselvier, (Eds.) New York.
67. Lao, L.L., Peppas, N.A., Boey, F.Y.C., Venkatraman, S.S. 2011. Modeling of drug release from bulk-degrading polymers. Int. J. Pharm. 418, 28-41.
68. Ebube, N.K., Jones, A.B. 2004. Sustained release of acetaminophen from a heterogeneous mixture of two hydrophilic non-ionic cellulose ether polymers, Int. J. Pharm. 272, 19-27.

69. Mitchell, K., Ford, J.L., Armstrong, D.J., Elliott, P.N.C., Rostron, C., Hogan, J.E. 1993. The influence of concentration on the release of drugs from gels and matrices containing Methocel. Int. J. Pharm. 100, 155-163.
70. Tiwari, S.B., Murthy, T.K., Pai, M.R., Mehta, P.R., Chowdary, P.B. 2003. Controlled release formulation of tramadol hydrochloride using hydrophilic and hydrophobic matrix system, AAPS Pharm. Sci. Tech. 4, 3, E31.
71. Imeson, A.P. 2000. Carrageenan. In G. O. Phillips & P. A. Williams (Eds.), Handbook of hydrocolloids, Cambridge, UK, pp. 87-102.
72. Picker, K.M., 1999. Matrix tablets of carrageenans. II. Release behavior and effect of added cations. Drug Dev. Ind. Pharm. 25, 339-346.
73. Schmidt, A.G., Wartewig, S., Picker, K.M. 2003. Potential of carrageenans to protect drugs from polymorphic transformation. Eur. J. Pharm. Biopharm. 56, 101-110.
74. Hoffman, A.S. 2002. Hydrogels for biomedical applications. Adv. Drug Deliver Rev. 54, 3-12.
75. Mangione, M.R., Giacomazza, D., Cavallaro, G., Bulone, D., Martorana, V., Biagio, P.L.S. 2007. Relation between structural and release properties in a polysaccharide gel system. Biophys Chem. 129, 18-22.
76. Park, S.Y., Lee, B.I., Jung, S.T., Park, H.J. 2001. Biopolymer composite films based on kappa-carrageenan and chitosan. Mater. Res. Bull. 36, 511-519.
77. Gupta, V.K., Hariharan, M., Wheatley, T.A., Price, J.C. 2001. Controlled-release tablets from carrageenans: effect of formulation, storage and dissolution factors. Eur. J. Pharm. Biopharm. 51, 241-248.
78. Tuleu, C., Khela, M.K., Evans, D.F., Jones, B.E., Nagata, S., Basit, A.W. 2007. A scintigraphic investigation of the disintegration behaviour of capsules in fasting subjects: A comparison of hypromellose capsules containing carrageenan as a gelling agent and standard gelatin capsules. Eur. J. Pharm. Sci. 30, 251-255.
79. Bornhoft, M., Thommes, M., Kleinebudde, P. 2005. Preliminary assessment of carrageenan as excipient for extrusion/spheronisation. Eur. J. Pharm. Biopharm. 59, 127-131.
80. Naim, S., Samuel, B., Chauhan, B., Paradkar, A. 2004. Effect of potassium chloride and cationic drug on swelling, erosion and release from kappa-carrageenan matrices. AAPS Pharm. Sci.Tech., 5, 1-8.
81. Leong, K.H., Chung, L.Y., Noordin, M.I., Mohamad, K., Nishikawa, M., Onuki, Y. 2011. Carboxymethylation of kappa-carrageenan for intestinal-targeted delivery of bioactive macromolecules. Carbohydr. Polym. 83, 1507-1515.

82. Bhardwaj, T.R., Kanwar, M., Lal, R., Gupta, A. 2000. Natural gums and modified natural gums as sustained-release carriers. Drug Dev. Ind. Pharm. 26, 1025-1038.
83. Christopher, M.J. 2003. Carrageenan-induced paw edema in the rat and mouse. Meth. Molecul. Biol. 225, 115-121.
84. Zhou, G., Sun, Y., Xin, H., Zhang, Y., Li, Z., Xu, Z. 2004. *In vivo* antitumor and immunomodulation activities of different molecular weight lambda-carrageenans from Chondrus ocellatus. Pharmacol. Res. 50, 47-53.
85. Caceres, P.J., Carlucci, M.J., Damonte, E.B., Matsuhiro, B., Zuniga, E.A. 2000. Carrageenans from Chilean samples of Stenogramme interrupta (Phyllophoraceae): structural analysis and biological activity. Phytochem. 53, 81-86.
86. Carlucci, M.J., Scolaro, L.A., Damonte, E.B. 1999. Inhibitory action of natural carrageenans on herpes simplex virus infection of mouse astrocytes. Chemotherapy 45, 429-436.
87. Graham, N.B. 1986. Hydrogel in controlled drug delivery. In E. Piskin, & A. S. Hoffman (Eds.), Polymeric material, Netherlands: Martinus Nijhoff Publishers.
88. Liu, J., Li, L., Cai, Y. 2006. Immobilization of campthothecin with surfactant into hydrogel for controlled drug release. Europ. Polym. J. 42, 1767-1774.
89. Hossain, K.S., Miyanaga, K., Maeda, H., Nemoto, N., 2001. Sol–gel transition behavior of pure -carrageenan in both salt-free and added salt states. Biomacromolecules 2, 442-449.
90. Changez, M., Burugapalli, K., Koul, V., Choudhary, V. 2003. The effect of composition of poly(acrylic acid)–gelatin hydrogel on gentamicin sulphate release: *in vitro*. Biomaterials 24, 527-536.
91. Sato, S., Kim, S.W. 1984. Macromolecular diffusion through polymer membranes. Int. J. Pharmacol. 22, 229-255.
92. Aranilla, C.T., Yoshii, F., dela Rosa, A.M., Makuuchi, K. 1999. Radiat. Phys. Chem., 55, 127.
93. Bonferoni, M.C., Rossi, S., Tamayo, M., Pedraz, J.L., Domiguez-Gil, A., Caramella, C. 1993. On the employment of κ-carrageenan in a matrix system. I. Sensitivity to dissolution medium and comparison with sodium carboxymethylcellulose and xanthan gum, J. Control. Rel. 26, 119-127.
94. Hao, J.S., Chan, L.W., Shen Heng, P.W.S. 2004. Complexation Between PVP and Gantrez Polymer and Its Effect on Release and Bioadhesive Properties of the Composite PVP/Gantrez Films, Pharm. Dev. Tech. 9, 379-386.
95. Guiseley, K.B. 1978. Modified kappa-carrageenan. US Patent Office, Patent No. 4096327.

96. Hosseinzadeh, H., Pourjavadi, A., Mahdavinia, G.R., Zohuriaan-Mehr, M.J. 2005. Salt- and pH-resisting collagen-based highly porous hydrogel. J. Bioact. Compat. Polym. 20, 475-90.
97. Silva, D.A., Paula de, R., Feitosa, J., Brito de, A., Maciel, J., Paula, H. 2004. Carboxymethylation of cashew tree exudate polysacharide. Carbohydr. Polym. 58, 163-171.
98. Yagi, T., Nagasawa, N., Iiroki, A., Tamada, M., Aranilla, C. 2010. Method of manufacturing gel using polysaccharides as raw materials. United States Patent Application Publication, 20,100,314,580 A1.
99. Leong, K.H., Chung, L.Y., Noordin, M.I., Mohamad, K., Nishikawa, M., Onuki, Y., et al. 2011. Carboxymethylation of kappa-carrageenan for intestinal-targeted delivery of bioactive macromolecules. Carbohydr. Polym. 83, 1507-1515.
100. Lawal, S., Lechner, M., Kulicke, W. 2008a. Single and multi-step carboxymethylation of water yam (Dioscorea alata) starch: Synthesis and characterization. Int. J. Biol. Macromol. 4, 429-435.

4 Konjac Polysaccharide for Drug Delivery

Wei Ha[1], Sheng Zhang[2], Yang Kang[1] and Bang-Jing Li[1]

[1]*Chengdu Institute of Biology, Chinese Academy of Science, Chengdu 610041, China.*

[2]*State Key Laboratory of Polymer Materials Engineering, Polymer Research Institute of Sichuan University, Chengdu 610065.*

4.1 Introduction

Natural polysaccharides and their derivatives have received more and more attention in the field of drug delivery systems due to their outstanding merits. The polysaccharides do hold advantages over the synthetic polymers, generally because they are nontoxic, less expensive, biodegradable, and freely available compared to their synthetic counterparts[1]. Therefore, polysaccharides seem to be the most promising materials in the preparation of nanometeric carriers for drug delivery systems. As a consequence of these potential applications, mountains of works dealing with the use of polysaccharides for drug delivery has remarkably increased over the last decades[2,3].

A very promising polysaccharide, which has been lately incorporated into the drug delivery fields is konjac glucomannan (KGM). KGM is a high-molecular weight, water soluble, non-ionic, natural polysaccharide isolated from the tubers of the *Amorphophallus* konjac plants which are the main crop in mountainous areas of China and Japan. It is mainly composed of a high-molecular weight glucomannan in which mannose and glucose units in a ratio of 1.6:1 are connected by β-(1→4) linkages. Compared to other polysaccharides, KGM has the ability to lower blood cholesterol and sugar level, helps with weight loss, promotes intestinal activity and reduces the risk of developing diabetes and heart disease. Moreover, KGM with its characteristics of low cost, good biocompatibility and biodegradability displays promising application in food and food additives, pharmaceuticals, biotechnology and the fine chemical industry.

In this chapter, we introduce the physicochemical and biological properties which are decisive for the exploitation of KGM and its derivates as a biomaterial for drug delivery. These properties include the structural organization, molecular weight, solubility, viscosity, gelling properties and degradation behavior. Next, we emphasize the recent development in designing novel drug delivery systems based on KGM and its derivates, four mechanisms are introduced to prepare KGM-based drug delivery materials, that is, blend with other materials, covalent crosslinking, polyelectrolyte complex, and the self-assembly. Finally, the specific and perspective applications in the future of KGM in the drug delivery field are discussed.

4.2 Konjac Glucomannan (KGM): Origin and Structure

Mannan is a homopolymer of (1→4) linked β-D-mannose residues which has been found in cell walls of some types of algae and used as storage carbohydrates in bulbs. Glucomannans are neutral polysaccharides of the mannan family produced by many plants, specifically in softwoods (hemicellulose), roots, tubers and many plant bulbs[4-10], which can be served as energy reserves and in some cases structural roles. Despite the variety of source, the most commonly used type of glucomananan is named konjac glucomannan (KGM), which is extracted from tubers of *Amorphohallus konjac* C. Koch (*Syn. Conophallus konjak Schott*)[11,12]. This member of the family *Araceae* is the source of the so-called "konjak flour" which is a popular article of food in China and Japan. The bulbs of three year old plants are cut into thin slices which are then dried and powdered to give konjak flour[13].

Up to now, the admissive structure of glucomannan is a random arrangement of [(1→4)-β-D-Glc-] and [(1→4)-β-D-Man-]. The ratio Glc/Man depends on the source and varies from 1/1 in iris bulbs to 1/5 in certain gymnosperms[14]. Table 4.1 summarized the different ratio Glc/Man of glucomannans from different source[15]. For example, glucomannan extracted from scotch pine and orchid tubers has ratios of 2.1:1 and 3.6:1, whereas glucomannan extracted from *Amorphohallus konjac* has a mole ratio of 1.6:1, respectively.

TABLE 4.1

Glucomannans from different sources[15]

Source	Mannose: glucose ratio	Degree of polymerisation
Eastern white pine (*Pinus strobes*)	3.8:1	90
Higanbana (*Lycoris radiata*)	4.0:1	730
Konjac (*Amorphophallus konjac*)	**1.6:1**	**>6000**
Lily (*Lilium auratum*)	2.7:1	220
Orchid (*Tubera salep*)	3.2:1	600
Ramie (*Boehmeria nivea*)	1.8:1	Not reported
Redwood (*Sequois sempervirens*)	4.2:1	60
Suisen (*Narcissus tazetta*)	1.5:1	Not reported

Constitution of KGM was the object of many investigations in 1920s and the classical strategy was hydrolysis approach[16,17]. Hydrolysis of methylated KGM was reported to give a mixture of 2,3,4-tri-O-methyl-D-glucose, 2,3,4-tri-O-methyl-D-mannose and 2,3,6–tri-O-methyl-D-mannose. Thus, the formula of KGM was proposed for the polysaccharide that has β-(1→4)-linked D-glucose and D-mannose residues as the main chain with branches joined through C-3 carbon of D-glucosyl and D-mannosyl residues[18,19].

As for the branching point, Kato and Matsuda isolated and identified several oligosaccharides containing β-D-mannopyranosyl-(1→3)-*O*-β-D-mannopyranosyl linkage[20]. Afterwards, Maeda *et al.* confirmed the branching structure through C-3 carbon of both D-glucosyl and D-mannosyl residues by analysis of the hydrolyzate of the permethylated glucomannan[21]. Combined the structures information of KGM over the years, they proposed the possible chemical structure of KGM[21], as shown in Figure 4.1. However, these values should be regarded cautiously due to the variability observed depending on the analytical procedures. For example, Katsuraya *et al.* investigated the constitution of KGM by methylation analysis and ^{13}C NMR spectroscopy, the results showed that KGM is composed of β-(1→4) linked D-glucosyl and D-mannosyl residues as the main chain with branches through β-(1→6)-glucosyl units. Degree of branching is about 8% and the ratio of terminal glucosyl units to mannosyl units is ca. 2[22].

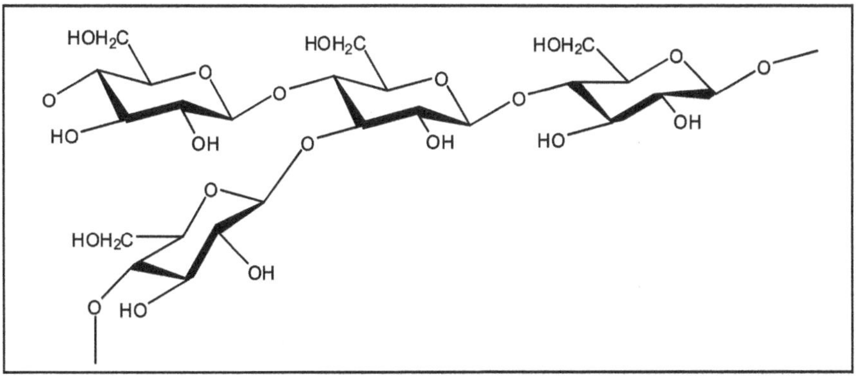

FIGURE 4.1 Possible structure of konjac glumannan.

4.3 Physicochemical Properties

4.3.1 Solubility

KGM is a kind of high molecular weight, water-soluble, nonionic polysaccharide. The component of KGM, unsubstituted linear β-(1→4) mannans and glucans (cellulose), are both insoluble in water owing primarily to inter-chain association through hydrogen bonding, yet KGM can be dissolved in water. This solubility may partly be attributed to long side chains of the glucomannan which can be served to hinder intermolecular association and enhance solubility[23,24]. Moreover, the degree of water solubility of KGM is particularly controlled by the presence of the acetyl units. More interesting, the presence of acetyl groups in the KGM has been demonstrated that can efficiently inhibit the formation of intramolecular hydrogen bonds leading to improve the KGM solubility[25]. However, after purification or drying processes, the solubility of KGM in water reduced with the increasing of the molecular weight due to the formation of strong hydrogen bonds. Great efforts have been devoted to improve the solubility of KGM. For example, considerable success in producing homogeneous solutions of KGM has been reported using "physical" methods whereby supramolecular aggregates are dispersed by increasing the energy of the component polymer chains. Such techniques include sonication, irradiation and the application of heat at elevated pressure[26]. Moreover, in recent years, a number of KGM derivatives have been synthesized in order to improve the KGM solubility. Besides water, solvents including isoamyl acetate and aqueous cadoxen (CdO/ethylenediamine) have been used to solubilize KGM[27].

4.3.2 Molecular Weight

The molecular weight (M_w) of KGM can be determined by light scattering, viscosimetry, gel permeation chomatography (GPC), capillary viscometry, photon correlation spectroscopy (PCS) and GPC coupled with multi angle laser light scattering (GPC-MALLS). One of the main problems in the determination of KGM M_w relies on its limited water solubility. In fact, some of M_w studies have been performed with KGM, which has been chemically modified in order to increase its solubility in water or other solvents. Torigata *et al.* studied the molecular weight of nitro-KGM in aqueous solution by light scattering and viscometry, the M_w were determined to be 2.7×10^5[28]. Water-soluble methylated KGM samples were prepared with a degree of substitution of about 0.45. The molecular weight was determined to be approximately 1.0×10^5[28]. Moreover, oxidation derivates of KGM were also prepared in order to increase its solubility in water to measure the M_w of KGM[29]. Such chemically modified strategy, however, time-consuming and may result in the degradation of the polymer. Solvents including isoamyl acetate and aqueous cadoxen have been used to solubilize KGM to measure the M_w by GPC approach[27]. However, it has been reported that KGM aggregates resist complete hydration even in 70% aqueous cadoxen. Viscometric measurements were also conducted on fractionated KGM samples, and the molecular parameters were similar to those obtained for several other polysaccharides. KGM formed very high-viscosity solutions with values higher than those of guar and locust bean gum, which indicated that the molecular mass of KGM was possibly higher assuming similar chain stiffness. Therefore, in order to measure the "true" molecular mass of KGM, the aqueous KGM solutions was disaggregated through controlled use of a microwave bomb, this approach facilitated reproducible molar mass distribution determination and alleviated the need for derivatization of polymer or the use of aggressive solvents. The M_w of KGM can also be characterized by aqueous GPC coupled with multi angle laser light scattering (GPC-MALLS). The results indicated that the most frequently used and commercially available KGM has a M_w in the range of 9.0 $(\pm 1.0) \times 10^5$ g/mol[26].

4.3.3 Gelation Properties

There are a number of parameters which affect the glumannan (GM) gelation behavior and the properties of the final gel structure. These parameters are the GM acetylation degree, the GM M_w, the temperature and also the concentration of both GM and the alkali involved in the gelation process[30].

The early studies identified the following gelation characteristics of KGM. The sol is transformed into a gel only in the presence of alkali and heat[31]. Kishida and Okimasu[32] found that the rheological properties of KGM vary depending on the gelling agent and the gelation proceeds more fully as temperature increases. Sakurada and Fuchino[33] reported that the gelation of the sol with alkali is accompanied by a conformational transformation of KGM, from an amorphous form to an ordered form.

Maekaji *et al.*[31] reported that the KGM would crystallize, in part, through a linkage such as hydrogen bonding to form a network structure once the low-molecular moiety was lost. The gelation process, however, does not proceed on standing, even though the moiety containing the C=O group is eliminated completely from native KGM. As far as the result is concerned, it is considered that the elimination of the moiety is necessary, but not sufficient for gelation. The further investigation explained that gelation of the sol of native KGM is initiated by stirring (all his work was done at 40 °C). In order to gelatinize the KGM sol, it is necessary to position the molecular or the segments of KGM so closely that the attractive forces among them can exert to fix the structure. The gelation process will proceed spontaneously if Brownian movement or the diffusion of the molecular occurs smoothly. And the mobility of molecular or segments may be restricted due to the high viscosity of the sol. Huang *et al.*[34] reported that in the presence of excessive alkali, the obtained KGM gel has smaller elastic modulus because of the gelation process of KGM were too fast and the saturated elastic modulus of gels strongly depends on the gelation rate. The further investigation demonstrated that the saturated elastic modulus of acetylated KGM gels were larger than that of native KGM, suggesting that the gelation rate is more important in the present case than the molecular weight difference (about twice) which is a determining factor for the saturated elastic modulus.

Up to this point it has been accepted that the KGM gel can be stabilized by hydrogen bond. Maekaji[31-37] reported that different reagents which can "peptize" the KGM gel under mild conditions. The brief process is as follows: The KGM gel was first formed in the presence of alkali and heat, then crushed into pieces, washed with distilled water and lyophilized. As many as 65 different reagents were tested at different concentrations. The lyophilized gel was hydrated with the peptizing agent. After holding for 20 hours at 5 °C, viscosities were obtained of the supernatant. The relative viscosity did not increase gradually with increasing concentration of peptizer but did abruptly distributed within a narrow range, i.e., 1.0-2.0 M of potassium thiocyanate (KSCN) or 3.0-4.0 M urea. Those reagents were classified into active, weak and inactive groups. Active reagents increased the viscosity of the supernatant solution significantly, indicating disruption of the gel and dissolution of the polymer. Weak reagents were able to swell the gel more than water, but did not increase viscosity. Finally, inactive reagents did not increase viscosity or swell the gel.

4.3.4 Derivates

KGM has been modified into various derivatives due to its good biocompatibility and biodegradable activity. A kind of KGM derivates with quaternary ammonium functionality was synthesized and optimal conditions for the quaternization were defined by Yua *et al.*[38]. Compared with the native KGM, quaternized konjac glucomannan (QKGM) displayed pronounced inhibitory effect on growth of four fungal strains and three bacterial strains. Such new properties opened a new possibility for commercial applications of KGM.

Acrylamide grafted konjac mannan (AKGM) was prepared by Xiao *et al.*[39] in the presence of Ce (IV) which can be served as initiator. The results showed that AKGM has better solubility in water, and the water solution of AKGM was more stable than KGM. Furthermore, the AKGM aqueous solution has a higher viscosity than that of KGM in most concentrations.

The esterification modified KGM is a kind of important derivates because of KGM has large number of hydroxyl functional group. The KGM derivates based on esterification modified strategy generally prepared under certain conditions with different acid or anhydride[40]. It including phosphate, sodium saliclate, benzoic acid, maleic anhydride, gallic acid, acetic acid and so on. After modified by esterification strategy with different moiety, the shear resistant, acid and alkali resistance of KGM were significantly improved.

The KGM acetate with high substitution degree is a kind of important derivate of KGM due to its typical viscoelastic property. The elastic ratio of KGM acetate is about 21.27%, and the stability of the workpiece size is very well. Furthermore it can be used as scale production by the means of conventional thermal plastic processing due to its excellent thermoplastic property. So it can be used to prepare the KGM degradable plastic. More interesting, the molecular weight of high substitution degree KGM acetate was significantly decreased compared with KGM[41]. Its decomposition temperature was 204.56 °C and the glass transition temperature was 128.49 °C. Melt shear viscosity η results indicated that the KGM acetate is very sensitive to the temperature.

Another valuable derivate of KGM is carboxymethyl konjac glucomannan (CKGM)[42]. Due to the introduction of hydrophilic carboxymethyl group, CKGM has excellent water solubility, and the resulting glue is more stable than native KGM. Furthermore, the film-forming property is highly improved. Therefore, CKGM can be applied for fruit preservation, printing and dyeing, etc. In addition, increasing the DS (degree of subsitution) resulted in higher water-absorbing and water-holding capacities. As a result it can be become water-holding and antimicrobial properties of food, it can be used to retain freshness and guarantee the quality with the stewing and baking food.

The product obtained from the deacetylation reaction of KGM in the presence of a trace alkali initiation has good water and heat resistance, biodegradablity and good tensile strength elongation at break[43]. Compared with unmodified film, the tensile and folding strength of the film prepared by modified KGM through deacetylation reaction directly both significantly improved and the uniformity of the film and surface was also improved to a certain degree. Moreover, chemical modification treatment by a mechanical force can effectively stripped of the acetyl group and chemical modification by the mechanical force could reduce the hydrosol thixotropy of KGM.

4.4 KGM based Drug Delivery Systems

KGM has a large number of reactive groups, a wide range of molecular weight and different chemical composition. Due to the presence of various derivable groups on molecular chains, KGM can be easily modified chemically and biochemically, resulting in many kinds of KGM derivatives. As natural biomaterials, KGM are highly stable, safe, non-toxic, hydrophilic and biodegradable. KGM have abundant resources in nature and low cost in their processing. In addition, KGM and its derivates have hydrophilic groups such as hydroxyl and amino groups, which could form non-covalent bonds with biological tissues, forming bioadhesion[44]. Particularly, KGM itself possess properties of anti-obesity activity, anti-hyperglycemic and hypercholesterolemia activities, laxative effect, prebiotic activity and anti-inflammatory activity[45]. Therefore, drug carriers made of KGM or its derivates not only could prolong the residence time and therefore increase the absorbance of loaded drugs, but also have variety potential health benefits. All these merits endow KGM a promising future as biomaterials. For the application of KGM for drug carriers, issues of safety, toxicity and availability are greatly simplified. In recent years, a large number of studies have been conducted on KGM and their derivatives for their potential application as drug delivery system.

4.4.1 Blend with Polymers

In the polymer science, natural polymer composite is still a current and important research branch[46]. Blending with polymer not only can increase the performance of a single polymer, compensate for the defects of single polymer's performance, but also preserve the original physical properties of the polymer. Meanwhile, it is a simple process and do have no pollution. So it is one of the most environmentally friendly ways to polymer modification[47].

KGM is a slow, controlled release drug excipient which has a lot of advantages compared to synthetic materials. However, KGM has the properties of poor water resistance and bad strength due to its good water solubility and large swelling ratio. In practice, KGM and the loaded drugs were always modified in drug delivery application. Therefore, constructed the polymer complex by KGM blending with other natural polymer is a more simple process which can increase the performance of KGM and compensate for the defects of KGM.

Yu *et al.*[48] utilized different proportion of KGM and poly(acrylic acid) mixture to get the blend membrane and matrix tablets. In the work, the existence of strong interactions between the carboxyl group of polyvinyl acetate and the hydroxyl group of KGM were proved by thermodynamic analysis. Moreover, such KGM and poly(acrylic acid) blend membrane was proved that has good controlled release effect by using ketoprofen as the model drug.

Hydrogels, the swellable polymeric materials, have been widely investigated as the carrier for drug delivery systems. These biomaterials have gained attention owing to their peculiar characteristics like swelling in aqueous medium, pH and temperature sensitivity or sensitivity towards other stimuli. Hydrogels being biocompatible materials have been recognized to function as drug protectors, especially for peptides and proteins, from *in vivo* environment. Also these swollen polymers are helpful as target carriers for bioactive drugs with tissue specificity. In consideration of excellent gelation ability and biocompatibility of KGM, plenty of biocompatible hydrogels containing KGM were prepared for drug carrier purposes[49].

Due to the fact that KGM is degraded by the enzymes secreted by the colonic bacteria, some authors have underlined the potential of KGM-based hydrogels for colonic drug delivery. Landin *et al.* (50,51) prepared thermoreversible gels composed of KGM and xanthan gum (XG). Rheological measurements of KGM and KGM/XG systems incubated with and without *Aspergillus niger* β-mannanase (used to mimic colonic enzymes) showed that KGM was degraded by the enzyme even when interacting with XG. Tablets with KGM/XG/sucrose matrices that varied in accordance with a simplex design and bore diltiazem as a typical highly soluble drug loading were prepared by wet granulation, and in most cases were found to possess satisfactory mechanical strength and exhibit slow, nearly zero-order drug release. Drug release from these tablets remained zero-order, but was accelerated (presumably due to degradation of KGM), in the presence of *A. niger* β-mannanase at concentrations equivalent to human colonic conditions.

4.4.2 Covalent Crosslinking

KGM and its derivates have many functional groups such as hydroxyl and amino groups, which could form covalent crosslinking with other polymers or biomacromolecules, forming novel stimuli-response hydrogels for site-specific drug delivery.

Xiao e*t al.* prepared a novel composite hydrogel by the use of dialdehyde KGM (DAK) as macromolecular cross-linking agent for chitosan (CS). KGM was initially oxidized by sodium periodate and then cross-linked to CS via imine bonds (–C=N–) to form the new DAK-CS hydrogels (Figure 4.2 (a) and (b)). The structure-function relationship associated with the mechanical, swelling, morphology and drug release properties could be modulated and correlated by varying the amount of oxidized KGM. The results of *in vitro* drug release illustrated that such hydrogels presented sustained release properties, which proposed to be applied as a drug controlled release system, especially as a vehicle for oral controlled release purposes or for colon targeting[52].

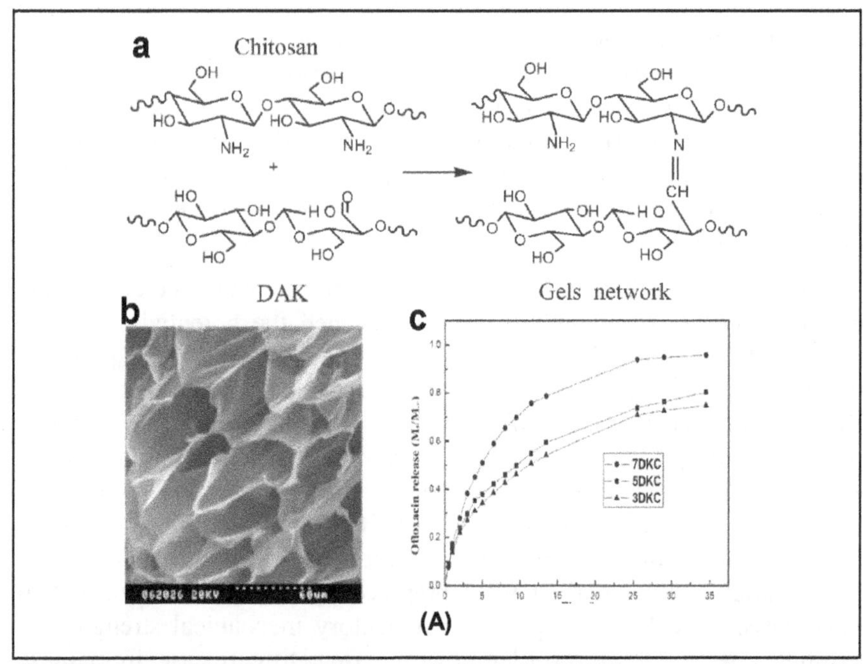

FIGURE 4.2 (A) (a) CS cross-linking with DAK via Schiff-base reaction; (b) SEM photographs of freeze-dried DAK-CS hydrogels; (c) Time dependence of ofloxacin release from the DAK-CS hydrogels at pH 7.4 (52).

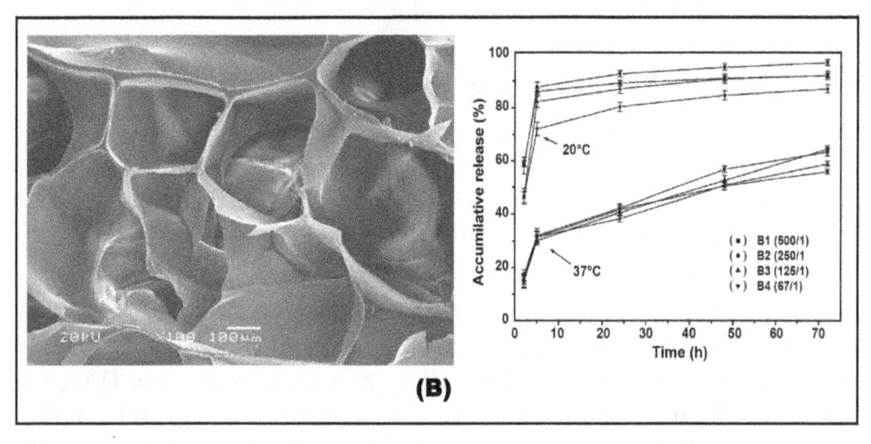

FIGURE 4.2 (B) SEM micrographs of lyophilized hydrogels (KGM/PNIPAM: 1/10) and in vitro release profile of BSA-loaded hydrogels (KGM/PNIPAM: 1/10) in PBS at 20 and 37°C. The data in brackets are the molar ratios of PNIPAM/BIS (53).

Xiong *et al.* prepared a kind of thermo-sensitive hydrogel. The hydrogel was made of KGM, copolymerized with N-isopropylacrylamide and crossilinked by *N,N*-methylene-bis-(acrylamide)[53]. The morphology of such hydrogels showed typical porous structure which is indispensable to allow for tissue growth, diffusion of drugs and nutrients. Moreover, this gels swelling ratio has possessed sensitive response to the environmental temperature. *In vitro* release behavior of model drug (bovine serum albumin) indicated that the release of the model drug was well controlled by the temperature.

Zhuo *et al.* synthesized a series of novel hydrogel system designed for colon-targeting drug delivery. The gels were composed of KGM, copolymerized with acrylic acid, and crosslinked by the aromatic azo agent[54] *N,N*-methyleen-bis-(acrylamide) (Figure 4.3) or bis (methacryloylamino)-azobenzene[55] (Figure 4.4), which have been investigated with regard to their biodegradability in the colon. The studies on the swelling behavior of those hydrogels revealed sensitive response to pH change. Furthermore, those gels retain the biodegradability and specificity to enzymatic degradation character of KGM. The model drug BSA or model drug 5-aminosalicylic acid (5-ASA) can be loaded by embedding into hydrogel network during the polymerization process, and the *in vitro* release behavior indicated that the release is controlled by swelling and degradation of the hydrogels.

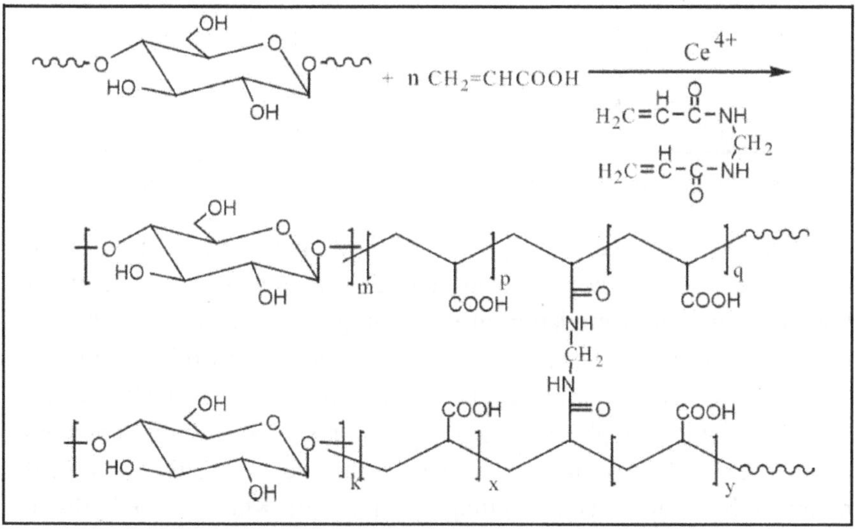

FIGURE 4.3 The synthetic route of KGM-*g*-acrylic acid hydrogles.

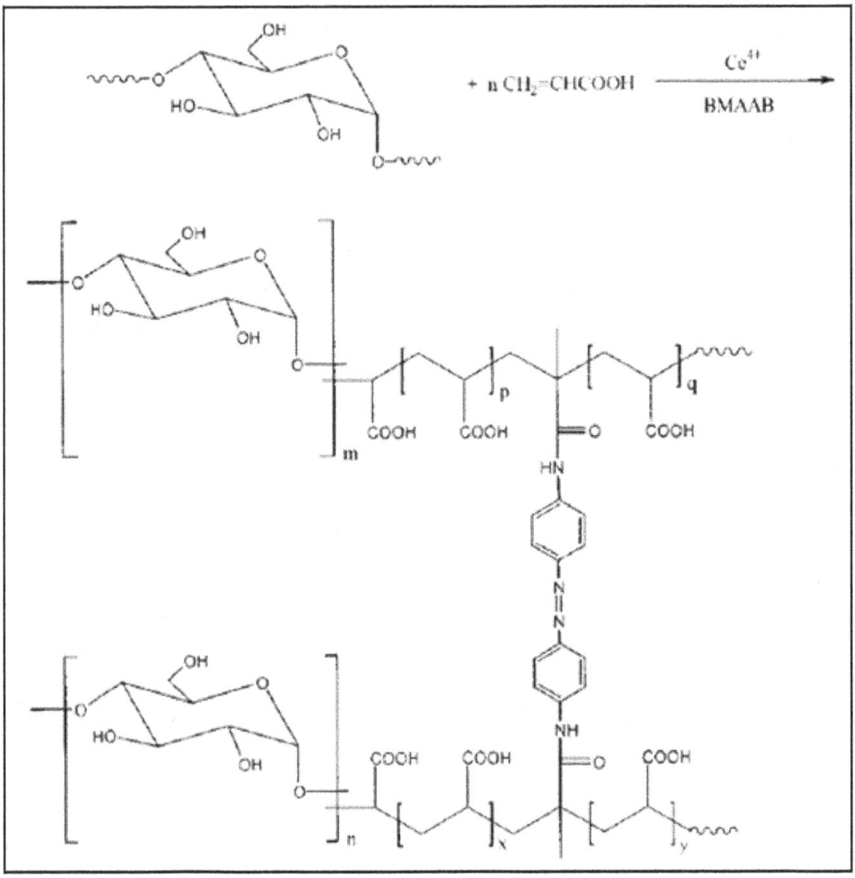

FIGURE 4.4 The synthetic route of KGM-*g*-AA hydrogles containing azo crosslinker.

Zhao *et al.* applied such PAA-KGM hydrogels for loading VB_{12}, the *in vitro* release behavior revealed that the release was also controlled by both swelling and degradation of the hydrogels[56]. Chen *et al.* reported a novel pH-sensitive interpenetrating polymer network (semi-IPN) hydrogels which prepared by KGM, poly(aspartic acid) (PAsp) with trisodium trimetaphophate as the cross-linking[57]. Such molar content of cross-linker and PAsp has a significant influence on swelling ratio of the obtained hydrogels. The swelling behavior of KGM-PAsp hydrogels revealed sensitive response properties to environmental pH change. In addition, the release behavior of 5-FU were also sensitive to the environmental pH (Figure 4.5).

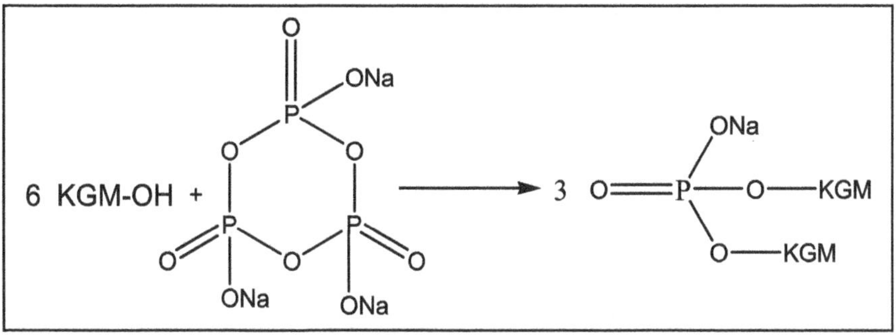

FIGURE 4.5 The crosslink mechanism.

He *et al.* prepared a kind of controlled release beads by using alginate, KGM and chitosan[58]. It was found that KGM could be contained within beads by faintness hydrogen binding and electrostatic interaction between alginate and KGM. Clear dents were observed on the surface of beads by SEM after introduction of KGM. The use of KGM could increase the payload of protein. Protein was released from alginate-chitosan beads within 1 h, while it was last from alginate-KGM-chitosan beads for 3 h. Furthermore, studies on water of hydration had shown that swelling ability of alginate-KGM-chitosan beads was higher than that of alginate-chitosan beads in acidic solution. Those results all indicated that the diffusion of protein was related to the excellent viscosity and swelling properties of KGM.

4.4.3 Polyelectrolyte Complex

Polyelectolyte can form polyelectrolyte complexation (PEC) with oppositely charged polymers by intermolecular electrostatic interaction. Polysaccharide-based PEC nanoparticles can be obtained by means of adjusting the MW of component polymers in a certain range. In theory, any polyelectrolyte could interact with polysaccharides to fabricate PEC nanoparticles (2). However, in practice, KGM is restricted to its water-soluble and non-ionic for fabricating PEC nanoparticles. In this sense, a kind of negative charged KGM derivates (carboxymethyl konjac glucomannan, CKGM) was synthesized. When treated with chloroacetic acid, the hydroxymethyl groups on the side chains of KGM are transformed into carboxymethyl groups to give CKGM. Due to its high viscosity, stability and emulsifiability in aqueous solution, CKGM shows promise in the pharmaceutical, chemical engineering, food, and papermaking industries as well as for environmental protection aspects. The structure of CKGM is shown in Figure 6. Chitosan (CS) is the only natural polycationic polysaccharide, therefore, the PEC formed between CKGM and CS for drug release application showed more valuable than other polyelectrolyte polymers in view of safety purpose. Du *et al.* prepared a kind of novel nanometer-sized particles driven by polyelectrolyte complexation between the hydrophilic CS and CKGM[59-61]. Such particles are shown to exist as discrete individual units with a circular shape consisting of a core and a semi-transparent shell which indicated that the nanoparticles posses a core (CKGM)-shell (CS) structure. Such CKGM-CS nanoparticles were prepared by dropwise addition of CKGM into CS solution. The size and size distributions of CKGM-CS nanoparticles were controlled by the concentrations of CKGM and CS (Table 4.2). These results may be due to an increased number of entwined molecule units at higher polymer concentrations, hence leading to the increase in nanoparticles size observed. Such CKGM-CS nanoparticles was further investigated as a protein carrier, the encapsulation efficiency of BSA varied from 30 to 45% depending on the initial loading concentration of BSA as well as the concentration of CS and CKGM employed for particle formation. Furthermore, the CKGM-CS nanoparticles showed excellent pH sensitivity which could be a suitable polymeric carrier for site-specific bioactive protein drug delivery.

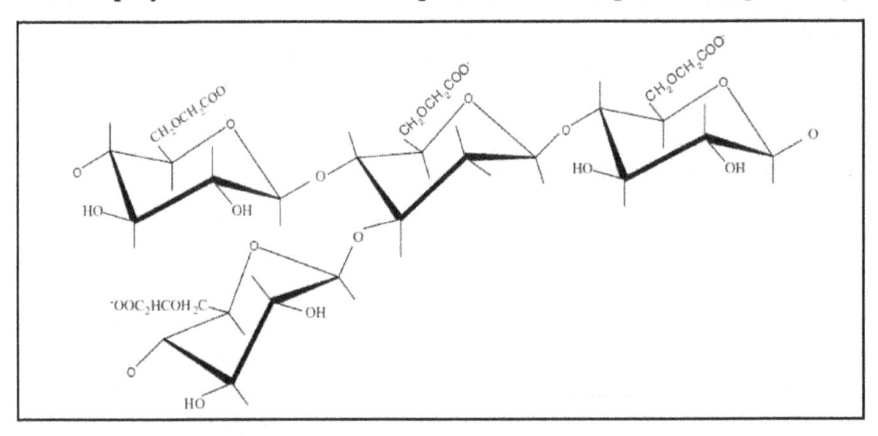

FIGURE 4.6 Structure of carboxymethyl Konjac glucomannan (CKGM).

L-asparaginase has been widely used to treat malignant tumors, particularly acute lymphoblastic leukemia (ALL). The growth of malignant tumor cells depends on the introduction of exogenous L-asparagine in most of the patients with ALL[62]. Tumor cells will languish without ingestion of exogenous L-asparagine whereas normal cells can synthesize L-asparagine themselves. L-asparaginase can effectively decrease the content of L-asparagine in blood by converting it to L-aspartic acid and ammonia. Although it has high therapeutic efficacy, there are still some problems that exist in the therapeutic field, such as immunological response and side effects (i.e., fever, skin rashes, allergic reactions and even anaphylactic shocks). In order to improve enzyme efficiency and reduce the immune response and toxicity, Li *et al.* prepared CKGM-CS nanoparticles as a novel biocompatible matrix system for L-asparaginase immobilization[63]. The preparation of the nanocapsules was completely conducted in water and the immobilized L-asparaginase maintained the original activity of the free enzyme. The encapsulation efficiency reached 68.0% when both the concentrations of CKGM and CS were 0.01% and the particle size was in a range 100-300 nm. Compared with the free L-asparaginase, the immobilized enzyme system showed significantly higher thermostability and had preferable resistance to acid and alkaline environments (Figure 4.7). This study illustrated that the nanocapsules have semi-permeability and can be used to immobilize thermal and pH-sensitive enzymes.

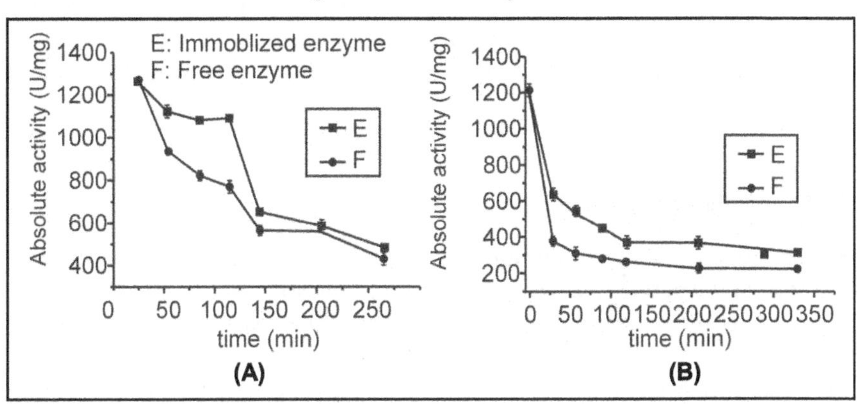

FIGURE 4.7 Thermal (a) and pH stability (b) of the free and immobilized L-asparaginase (pH 6) (63).

TABLE 4.2

Effect of the concentration of CKGM and CS on the size of nanoparticles (59-61).

CKGM	CS					
	The mean sizes and size distributions					
	0.01%	0.02%	0.04%	0.06%	0.08%	0.1%
0.01%	48.5	68.5	68.7	217.4	271.8	291.4
0.02%	89.4	121.4	209.1	329.4	366.0	548.6
0.04%	101.2	120.9	256.2	341.1	425.5	672.1

TABLE 4.2 ***Contd...***

CKGM	CS					
	The mean sizes and size distributions					
	0.01%	0.02%	0.04%	0.06%	0.08%	0.1%
0.06%	142.6	166.2	320.6	486.2	546.2	891.9
0.08%	198.7	231.5	354.0	511.6	654.6	1034.4
0.1%	305.0	333.9	402.7	518.3	836.1	1198.7

Noteworthy is that, the category of ionic-KGM was only negative derivates hitherto reported, therefore, the category of polyelectrolyte complex containing KGM derivatives for drug delivery will continuously increase because:

1. Electropositive KGM derivates will endow more probability for constructing drug carriers based on polyelectrolyte complex with other negative polymers;
2. The drug carriers formed by complexion between KGM derivates and biomacromolecular, such as peptide, protein, DNA and RNA is a special type of delivery system, in particular, will be studied ulteriorly.

4.3 Self-Assembly of KGM Derivates

Self-assembly is the autonomous organization of components into patterns or structures without human intervention. The concept of self-assembly is used increasingly in many disciplines, with a different flavor and emphasis in each. The drug carriers based on self-assembly strategy attracted researchers great interest in recently years due to its simple preparation process and various morphology and function.

4.3.1 Hydrophobically Modified KGM

When hydrophilic polymeric chains are grafted with hydrophobic segments, amphiphilic copolymers are synthesized. Within an aqueous environment, polymeric amphiphiles spontaneously form micelles or micelle-like aggregates via undergoing intra- or intermolecular associations between hydrophobic moieties, primarily to minimize interfacial free energy. These polymeric micelles exhibit unique characteristics, depending on hydrophilic/hydrodynamic radius with core-shell structure. In particular, polymeric micelles have been recognized as a promising drug carrier, since their hydrophobic domain, surrounded by hydrophobic drugs (2).

Li *et al.* designed and synthesized a cholesterol-modified CKGM (CHCKGM) conjugate by introducing cholesterol molecules into the glucose units of CKGM[64]. By varying the feed ratio of cholesterol, a series of CHCKGM conjugates were prepared with different grafting ratios. The critical aggregation concentration (cac) values of conjugates are smaller than 5.89×10^{-3} mg/mL, indicating that the cholesterol residue is very effective to form aggregates (Figure 4.8). Those self-aggregates exhibit a pH-and ionic strength-dependent properties causing a considerable increase (decrease) of their radius. Hydrophobic drug etoposide was successfully entrapped into the CHCKGM nanoparticles and its release behavior *in vitro* exhibited a sustained release (Figure 4.9).

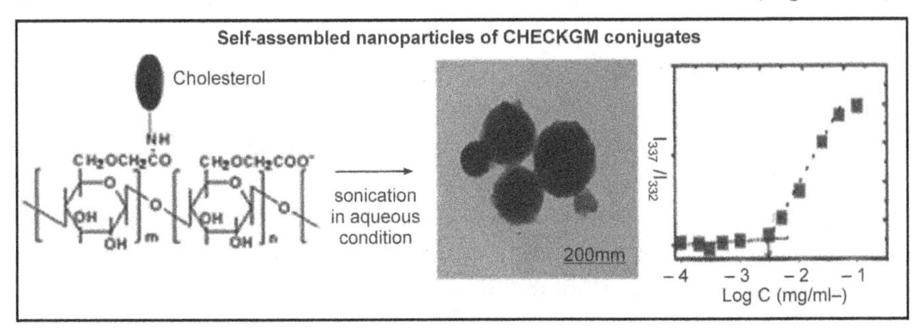

FIGURE 4.8 TEM micrograph and cac values of self-assembled nanoparticles of CHCKGM conjugates[64].

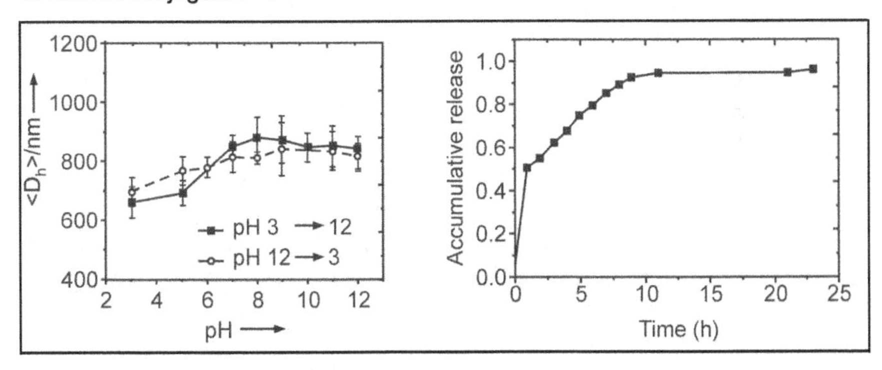

FIGURE 4.9 Average hydrodynamic diameter <D_h> of CHCKGM self-aggregated spheres as a function of pH and Etoposide release from etoposide/CHCKGM nanoparticles at 30 °C in PBS (pH 7.4)[64].

In recent years, numerous studies have been carried out to investigate the synthesis and the application of hydrophobic molecules modified polysaccharide self-aggregate nanoparticles as drug delivery systems. However, the synthesis and characterization of modified CKGM with hydrophobic moieties and further using this amphiphile to form self-aggregated nanoparticles for drug delivery rarely have been reported. In fact, there are many hydrophobic molecules, in particular, hydrophobic drug, can be used to modify KGM for drug-related applications.

4.3.2 Polyethylene Glycol Modified KGM

Polymeric hollow spheres may be ideal structures for the encapsulation of large quantities of guest molecules or for large-sized guests within the core domain of the polymer. Activity in this area of research has increased since rigid-coil copolymer self-assembly and the direct formation of hollow spheres in a selective solvent was described[65,66]. This novel approach has aroused recent investigations in Li's group and other group using rigid/coil systems to construct hollow spheres for drug carrier application. Several reports have demonstrated that the use of hydrogen-bonding interactions between rod-like and coil-like polymers which can directly self-assemble into hollow spheres. These self-assembled hollow spheres, however, lack biocompatibility and biodegradability because the rod-like blocks were formed by synthetic methods. For biomedical and drug deliver applications, the self-assembled hollow nanospheres that are biocompatible and non-toxic would be most desirable.

Li *et al.* developed a novel approach to prepare rod-coil complexes and hollow nanospheres by self-assembly of CKGM-*g*-poly(ethylene glycol) and α-CD, in which rod-like segments were formed by inclusion complex between α-CD and graft PEG (Figure 4.10). These hollow nanospheres showed semi-permeability and the encapsulated enzyme glucose oxidase (GOX) exhibited significantly higher stability and sustained enzyme activity over a greater range of pH and temperature environments when compared with the free enzyme (Figure 4.11)[67,68]. They envision that the bio-compatibility of CKGM-*g*-PEG/α-CD hollow nanospheres is suitable for certain applications in protein, enzyme, DNA or RNA encapsulation and delivery, bioreactors, or as biosensor devices.

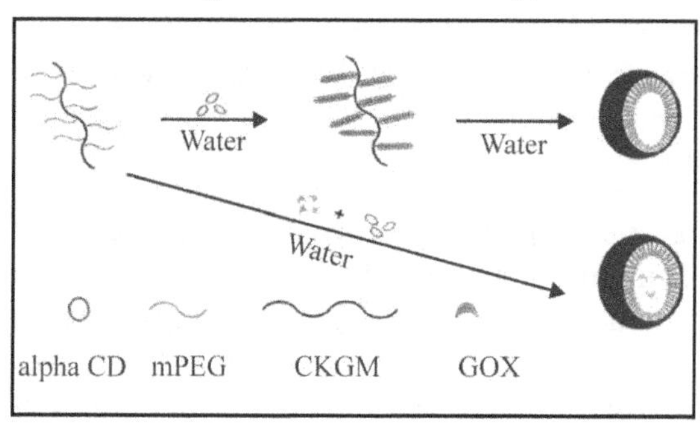

FIGURE 4.10 Construction of CKGM-*g*-PEG/α-CD hollow nanospheres and the process of enzyme encapsulation[68].

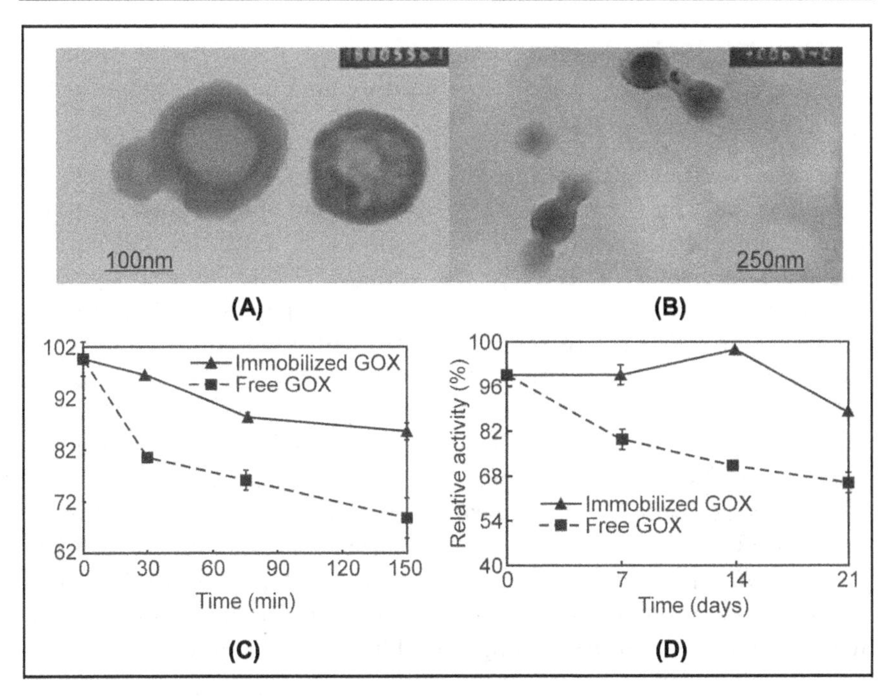

FIGURE 4.11 TEM micrograph of (a) CKGM-*g*-PEG/α-CD hollow nanospheres (b) CKGM-*g*-PEG/α-CD hollow particles with GOX. (c) Thermostability of the immobilized and free GOX at 47 ℃ (d) Storage stabilities of the immobilized and free GOX[68].

4.3.3 Others

Peptide-based hydrogels formed through molecular self-assembly are an important class of biomedical materials. The self-assembling peptide hydrogels have many obvious advantages over other traditional hydrogel systems[69]. For example,

(a) The peptides that construct the hydrogels can often be degraded *in vivo*, and the resulting products are nontoxic;

(b) The hydrogels are spontaneously formed without using harmful chemicals such as cross-linkers;

(c) The spontaneous process allows for soulution-gel transformation *in vivo* by injecting peptide solutions at specific locations, and it also enables a facile incorporation of cell-specific bioactive moieties into hydrogels;

(d) The peptide building blocks represent a variety of chemical groups that make hydrogels be easily modified with chemical and biological moieties.

These desirable characteristics have attracted great interest in designing and constructing peptide hydrogels. He *et al.* prepared a novel peptide-polysaccharide hybrid hydrogel as a potential carrier for sustained delivery of hydrophobic drugs[69]. The hybrid hydrogel composed of Fmoc-diphenylalanine (Fmoc-FF) peptide and KGM was prepared through molecular self-assembly of Fmoc-FF in the KGM solution. Such hybrid hydrogel exhibited a highly hydrated, rigid and nanofibrous gel network in which self-assembled peptide nanofibers were interwoven with the KGM chains. Moreover, docetaxel was chosen as a model of hydrophobic drugs and the *in vitro* release behavior showed sustained and controlled drug release by varying the KGM concentration, molecular weight, aging time or β-mannanse concentration. Those works not only provide a new strategy for fabricating peptide-KGM hybrid hydrogel as a sustained-release drug carrier, but also open an avenue for the design of new self-assembling hybrid hydrogels based on KGM.

Perspective

As reviewed above, various drug delivery systems based on KGM or its derivates have been prepared. However, compared with other polysaccharide such as chitosan, alginate, the studies is not wide and deep enough. It can be predicted that, more novel drug carrier systems containing KGM will emerge in the near future. Until now, those drug carriers based on KGM are only investigated in terms of their physicochemical properties, drug-loading ability, *in vitro* toxicity, and comparatively simple *in vivo* tests. As an important part of polysaccharide family, the interaction between KGM and human cells, organs, tissues, bio-molecules, the effect on human's metabolism, the synergy effect of encapsulated drug and KGM itself for some special disease and the other wide application for drug delivery, etc. await for further study.

Reference

1. Sinha, V.R.; Kumria, R. Polysaccharides in colon-specific drug delivery, Int. J. Pharm. 2001, 224, 19-38.
2. Liu, Z.H.; Jiao, Y.P.; Wang, Y.F.; Zhou, C.R.; Zhang, Z.Y. Polysaccharides-based nanoparticles as drug delivery systems. Adv. Drug Deliv. Rev. 2008, 60, 1650-1662.

3. Janes, K.A.; Calvo, P.; Alonso, M.J. Polysaccharide colloidal particles as delivery systems for macromolecules, Adv. Drug Deliv. Rev. 2001, 47, 83-97.
4. Bouveng, H.-O.; Iwasari, T.; Lindberg, B.; Meier, H. Studies on glucomannans from Norwegian spruce. 4. Enzymic hydrolysis, Acta Chem. Scand. 1963, 17, 1796-1797.
5. Timell, T.-E. Wood hemicelluloses: Part I, Adv. Carbohydr. Chem. 1964, 19, 247-302.
6. Nozawa, Y.; Hiraguri, Y.; Ito, Y. Studies on the acid stability of neutral monosaccharides by gas chromatography with reference to the analysis of sugar components in the polysaccharides, J. Chromatogr. 1969, 45, 244-249.
7. Franz, G.; Biosynthesis of salep mannan, Phytochemistry 1973, 10, 2369-2373.
8. Kenne, L.; Rosell, K.-G.; Svensson, S. Distribution of the O-acetyl groups in pine glucomannan, Carbohydr. Res. 1975, 1, 69-76.
9. Ishrud, O.; Zahid, M.; Viqar, U.-A.; Pan, Y.-J. Isolation and structure analysis of a glucomannan from the seeds of Libian dates, Agric. Food Chem. 2001, 8, 3772-3774.
10. Rhodes, D.I.; Stone, B.-A. Proteins in walls of wheat aleurone cells, J. Cereal Sci. 2002, 1, 83-101.
11. Xiao, C.; Gao, S.; Zhang, L. Blend films from konjac glucomannan and sodium alginate solutions and their preservative effect, J. Appl. Polym. Sci. 2000, 3, 617-626.
12. Xiao, C.; Gao, S.; Wang, H.; Zhang, L. Blend films from chitosan and konjac glucomannan solutions, J. Appl. Polym. Sci. 2000, 4, 509-515.
13. Smith, B.F.; Srivastava, H.C. Constitutional studies on the glucomannan of konjak flour, J. Am. Chem. Soc. 1959, 81, 1715-1718.
14. Rinaudo, M. Main properties and current applications of some polysaccharides as biomaterials, Polym. Int. 2008, 57, 397-430.
15. AL-Ghazzewi, F.H.; Khanna, S.; Tester, R.F.; Piggott, J. The potential use of hydrolysed konjac glucomannan as a prebiotic, J. Sci. Food Agric. 2007, 87, 1758-1766.
16. Maeda, M.; Shimahara, H.; Sugiyama N., Detailed examination of the branched structure of konjac glucomannan. Agric. Biol. Chem. 1980, 44, 245-252.
17. T. Ohtsuki, Studien ueber das Konjacmannan. Acta Phytochimica (Japan), 1929, 4, 1-39.
18. Nishida, K.; Hashima, H. Chemische untersuchungen ueber das glukomannan aus konjac. Journal of the Department of Agriculture, 1930, 2, 277-360.

19. Smith, F.; Srivastava, H. C. Constitution studies on the glucomannan of konjac flower. J. Am. Chem. Soc. 1959, 81, 1715-1718.
20. Kato, K.; Matsuda, K. Studies on the chemical structure of konjac mannan. Isolation of oligosaccharides corresponding to the branching point of konjac mannan. Agric. Biol. Chem. 1973, 37, 2045-2051.
21. Maeda, M.; Shimahara, H.; Sugiyama, N. Detailed examination of the branched structure of konjac glucomannan. Agric. Biol. Chem. 1980, 44, 245-252.
22. Katsuraya, K.; Okuyama, K.; Hatanaka, K.; Oshima, R.; Sato, T.; Matsuzaki, K. Constitution of konjac glucomannan: chemical analysis and ^{13}C NMR spectroscopy, Carbohydr. Polym. 2003, 53, 183-189.
23. Williams, M.A.K.; Foster, T.J.; Martin, D.R.; Norton, I.T. A molecular description of the gelation mechanism of konjac mannan, Biomacromolecules, 2000, 1, 440-450.
24. Hwang, J.; Kokini, J.L. Structure and rheological function of side branches of carbohydrate polymers, J. Text. Stud. 1991, 22, 123-167.
25. Gao, S.; Nishinari, K. Effect of deacetylation rate on gelation kinetics of glucomannan, Colloids Surf. B Biointerfaces 2004, 38, 241-249.
26. Ratcliffe I.; Williams P.A.; Viebke C.; Meadows J. Physicochemical characterization of konjac glucomannan, Biomacromolecules 2005, 6, 1977-1986.
27. Kohyama, K.; Sano, Y.; Nishinari, K. A mixed system composed of different molecular weights konjac glucomannan and k-carrageenan. II. Molecular weight dependence of viscoelasticity and thermal properties, Food Hydrocolloids 1996, 10, 229-238.
28. Dave, D.; McCarthy, S. P. Review of konjac glucomannan, J. Environ. Polym. Degr. 1997, 5, 237-241.
29. Benincasa, M. A.; Cartoni, G.; Delle Fratte, C. Flow field-flow fractionation and characterization of ionic and neutral polysaccharides of vegetable and microbial origin, J. Chromatogr. A 2002, 967, 219-234.
30. Alonso-Sande, M.; Teijeiro-Osorio, D.; Remunan-Lopez, C.; Alonso, M.J. Glucomannan, a promising polysaccharide for biopharmaceutical purposes, Eur. J. Pharm. Biopharm. 2009, 72, 453-462.
31. Maekaji, K.; Mechanism of Gelation of Konjac Mannan, Agr. Biol. Chem. Tokyo, 1974, 38, 315-321.
32. Kishida, N.; Okimasu, S. Preparation of water-soluble methyl konjac glucomannan, Agr. Biol. Chem Tokyo, 1978, 42, 669-670.
33. Sakurada, I.; Hutino, K. X-ray diagrams of glucommannan fibre, Phys. Chem.B-Chem. E, 1933, 21, 18-24.

34. Huang, L.; Takahashi, R.; Kobayashi, S.; Kawase, T.; Nishinari, K. Gelation behavior of native and acetylated konjac glucomannan, Biomacromolecules, 2002, 3, 1296-1303.
35. Ohta, Y.; Maekaji, K. Preparation of konjac mannan gel, J. Agr. Chem Soc. Jpn. 1980, 54, 741-746.
36. Maekaji, K. Kinetic Study on Gelation of Konjac Mannan .1. Method for Measurement and Kinetic-Analysis of Gelation Process of Konjac Mannan, J. Agr Chem. Soc. Jpn. 1978, 52, 251-257.
37. Maekaji, K.; Kawamura, D. Variation in Rheological Properties of Konjac Gel with the Corm Cultivar, Agr. Biol. Chem. Tokyo 1985, 49, 2483-2484.
38. Yu, H.Q.; Huang, Y.H.; Ying, H.; Xiao, C.B. Preparation and characterization of a quaternary ammonium derivative of konjac glucomannan, Carbohydr. Polym. 2007, 69, 29-40.
39. Xiao, C.B.; Gao, S.J.; Li, G.R.; Zhang, Q.C. Preparation of konjac glueomannan and acrylamide grafted konjac glucomannan. Wuhan University Journal of Natural Science 1999, 4, 459-462.
40. Chen, L.G. Research progress in the modification of konjac glucomannan. J. Anhui Agri. Sci. 2008, 36, 6157-6160.
41. Lin, X.Y.; Wu, Q.; Luo, X.G.; Zhang, S.; Luo, X.Q. Synthesis of konjac glucomannan acetate with high degree of substitution and of its rheological and thermoplastic properties. Acta Polymerica Sinica 2010, 7, 884-891.
42. Wang, C.; Li, B.; Xie, B.J.; Chen, J.X. Improvement of the preparation method for carboxy methylation of konjac glucomannan. Transactions of The Chinese Society of Agricultural Engineering 2005, 21, 140-145.
43. Pan, T.T.; Chen, L.P.; Wang, B.H. Progress of konjac glucomannan modification. Journal of the Chinese Cereals and Oils Association 2012, 27, 124-128.
44. Nakano, M.; Takikawa, K.; Arita, T. Release characteristics of dispersed in konjac gels, J. Biomed. Mater. Res. 1979, 13, 811-819.
45. Chua, M.; Baldwin, T.C.; Hoching T.J.; Chan, K.; Traditional uses and potential health benefits of Amorphophallus konjac K. Koch ex N.E.Br, J. Ethnopharmacol. 2010, 128, 268-278.
46. Gao, S.; Zhang, L. Semi-interpenetrating polymer networks from castor oil-based polyurethane and nitrokonjac glucomannan. J. Appl. Polym Sci. 2001, 81, 2076-2773.
47. Xiao, C.B.; Liu, H.J.; Gao, S.J. Charaterization of poly(vinyl alcohol)-konjac glucomannan blend film. J Macromol. Sci-Pure. Appl. Chem. 2000, 37, 1009-1021.

48. Yu, H.Q.; Huang, A.B.; Xiao, C.B. Characteristics of konjac glucomannan and poly (acrylic acid) blend films for controlled drug release. J. Appl. Polym. Sci. 2006, 100, 1561-1570.
49. Amin, S.; Rajabnezhad, S.; Kohli, K. Hydrogels as petential drug delivery systems, Scientific research and essay 2009, 3, 1175-1183.
50. Mancenido, F.A.; Landin, M.; Lacik, I.; Martínez-Pacheco, R. Konjac glucomannan and konjac glucomannan/Xannan gum mixtures as excipients for controlled drug delivery systems. Diffusion of small drugs, Int. J. Pharm. 2008, 349, 11-18.
51. Mancenido, F.A.; Landin, M.; Martínez-Pacheco, R. Konjac glucomannan/ xannan gum enzyme sensitive binary mixtures for colonic drug delivery, Eur. J. Pharm. Biopharm. 2008, 69, 573-581.
52. Yu, H.Q.; LU, J.; Xiao, C.B. Preparation and properties of novel hydrogels from oxidized konjac glucomannan cross-linked chitosan for in vitro drug delivery, Macromol. Biosci. 2007, 7, 1100-1111.
53. Xiong, Z.C.; Chen H.C.; Huang, X.C.; Xu, L.A.; Zhang, L.F.; Xiong, C.D. Preparation and properties of thermo-sensitive hydrogels of konjac glucomannan grafted N-isopropylacrylamide for controlled drug delivery, Iran. Polym. J. 2007, 16, 425-431.
54. Chen, L.G.; Liu, Z.L.; Zhuo, R.X. Synthesis and properties of degradable hydrogels of konjac glucomannan grafted acrylic acid for colon-specific drug delivery, Polymer 2005, 46, 6274-6281.
55. Liu, Z.L.; Hu, H.; Zhuo, R.X. Konjac glucomannan-graft-acrylic acid hydrogels containing azo crosslinker for colon-specific delivery, Journal of polymer science: polymer chemistry 2004, 42, 4370-4378.
56. Wen, X.; Cao, X.L.; Yin, Z.H.; Wang, T.; Zhao, C.S. Preparation and characterization of konjac glucomannan-poly(acrylic acid) IPN hydrogels for controlled release, Carbohydr. Polym. 2009, 78, 193-198.
57. Liu, C.H.; Chen, Y.Q.; Chen, J.G. Synthesis and characteristics of pH-sensitive semi-interpenetrating polymer network hydrogels based on konjac glucomannan and poly(aspartic acid) for in vitro drug delivery, Carbohydr. Polym. 2010, 79, 500-506.
58. Wang, K.; He, Z.M. Alginate-konjac glucomannan-chitosan beads as controlled release matrix, Int. J. Pharm. 2002, 244, 117-126.
59. Du, J.; Sun, R.; Zhang, S.; Govender, T.; Zhang, L.F.; Xiong, C.D.; Peng, Y.X. Novel polyelectrolyte carboxymethyl konjac glucomannan-chitosan nanoparticles for drug delivery, Macromol. Rapid. Commun. 2004, 25, 954-958.

60. Du, J.; Sun, R.; Zhang, S.; Zhang, L.F.; Xiong, C.D.; Peng, Y.X. Novel polyelectrolyte carboxymethyl konjac glucomannan-chitosan nanoparticles for drug delivery. I. Physicochemical characterization of the carboxymethyl konjac glucomannan-chitosan nanoparticles, Inc. Biopolymers, 2005, 78, 1-8.
61. Du, J.; Dai, J.; Liu, J.L.; Dankovich, T. Novel pH-sensitive polyelectrolyte carboxymethyl konjac glucomannan-chitosan beads as drug carriers, React. Funct. Polym. 2006, 66, 1055-1061.
62. Campbell, H.A.; Mashburn, L.T.; Boyse, E.A.; Old, L.J. Two l-asparaginase from Escherichia coli B: their separation, purification, and antitumor activity. Biochemistry, 1967, 6, 721-730.
63. Wang, R.; Xia, B.; Li, B.J.; Peng, S.L.; Ding, L.S.; Zhang, S. Semi-permeable nanocapsules of konjac glucomannan-chitosan for enzyme immobilization, Int. J. Pharm. 2008, 364, 102-107.
64. Ha, W.; Wu, Hao, Wang, X.L.; Peng, S.L.; Ding, L.S.; Zhang, S.; Li, B.J. Self-aggregates of cholesterol-modified carboxymethyl konjac glucomannan conjugate: Preparation, characterization, and preliminary assessment as a carrier of etoposide, Carbohydr. Polym. 2011, 86, 513-519.
65. Jenekhe, S. A.; Chen, X. L. Self-assembled aggregates of rod–coil block copolymers and their solubilization and encapsulation of fullerenes. Science 1998, 279, 1903-1907.
66. Jenekhe, S. A.; Chen, X. L. Self-assembly of ordered microporous materials from rod–coil block copolymers. Science 1999, 283, 372-375.
67. Xia, B.; Ha, W.; Meng, X.W.; Govender, T.; Peng, S.L.; Ding, L.S.; Li, B.J.; Zhang, S. Preparation and characterization of a poly(ethylene glycol) grafted carboxymethyl konjac glucomannan copolymer, Carbohydr. Polym. 2010, 79, 648-654.
68. Li, Q.; Xia, B.; Branham, M.; Ha, W.; Wu, Hao, Peng, S.L.; Ding, L.S.; Li, B.J.; Zhang, S. Self-assembly of carboxymethyl konjac glucomannan-g-poly(ethylene glycol) and (α-cycloxextrin) to biocompatible hollow nanospheres for glucose oxidase encapsulation, Carbohydr. Polym. 2011, 86, 120-126.
69. Huang, R.L.; Qi, W.; Feng, L.B.; Su, R.X.; He, Z.M. Self-assembling peptide-polysaccharide hybrid hydrogel as a potential carrier for drug delivery, Soft Matter 2011, 7, 6222-6230.

5 Chitosan and Modified Chitosans for Drug Delivery Application

Adeleke Omodunbi Ashogbon

Department of Chemistry and Industrial Chemistry, Adekunle Ajasin University, Akungba-Akoko, Ondo State, Nigeria.

5.1 Introduction

Chitin and chitosan are aminoglucopyrans composed of N-acetylglucosamine and glucosamine residue. CS is a natural, cheap, non-toxic cationic, biodegradable, biocompatible and renewable polymer obtained from chitin by alkaline de-acetylation. Chitin is the most abundant biopolymer in nature after cellulose, and it is a linear polysaccharide that is widely distributed in nature. Chitin is the main component of exoskeletons of crustaceans and insect as well as of cell wall of some bacteria and fungi. There are some resemblance between chitin and cellulose, both are glucose-based linear polysaccharide. CS differs from cellulose at the C-2 carbon by having an acetamide residue in place of a hydroxyl group.

CS is semi-crystalline and the degree of crystallinity is a function of the degree of deacetylation. It is insoluble at neutral and alkaline pH, but forms water-soluble salts with inorganic and organic acids among which are acetic, lactic, hydrochloric and glutamic acids. CS has been extensively evaluated for its mucoadhesive and absorption enhancement properties. The positive charge on the CS molecule gained by acidic environment in which it is soluble seems to be important for absorption enhancement. However, CS is not soluble in medium except below pH 5.6. This limits its use as permeation enhancer in body compartments where pH is high. That is the reason for the preparation of modified CS or CS derivatives with increase solubility, especially at neutral and basic pH values.

CS has a primary amino group, and a primary and secondary free hydroxyl groups. The strong functionality of CS due to these groups gives it a considerable opportunity of chemical modification. The purpose of this chapter is to discuss the properties of CS and modified CS as associated with drug delivery systems. In a nutshell, the chapter consists of: chemistry of chitosan relevant to its chemical modification and drug delivery; chemically modified CS for drug delivery systems; importance of mucoadhesive properties of CS in drug delivery systems and usage of CS and modified CS in drug delivery systems.

5.2 Chemistry of Chitosan Relevant to its Chemical Modification and Drug Delivery

Chitosan (CS) is the most important derivative obtained from chitin. Chitin is the second most ubiquitous natural polysaccharide after cellulose on earth. CS is a partially deacetylated polymer obtained from the alkaline deacetylation of chitin which is a glucose-based linear polysaccharide and the principal component of exoskeletons of crustaceans and insects, as well as of the cell walls of some bacteria and fungi. CS is semi-crystalline and its non-toxicity, biodegradability, biocompatibility, eco-friendly, cheapness and abundant had been widely reported in the literature[1-3].

CS is the form of chitin [β (1→4)-linked 2-acetamido-2-deoxy-β-D-glucose or N-acetylglucosamine] which has been deacetylated to at least 50% of the free amine form, which has a heterogeneous chemical structure made up of both 1-4 linked 2-acetamido-2-deoxy-β-D-glucopyranose (N-acetylglucosamine) as well as 2-amino-2-deoxy-β-D-glucopyranose (N-glucosamine). So, CS is not a constitutionally defined compound. In CS, the N-acetyl group is replaced either fully or partially by NH_2, therefore the degree of acetylation (DA) can vary from DA = 0 (fully deacetylated to DA = 1 (fully acetylated, i.e., chitin). Commercially available CS has an average molecular weight ranging between 3800 and 20,000 Daltons and is 66 to 95% deacetylated[1].

The basicity of CS due to the presence of primary amino group resulted into some unique properties like polyoxysalt formation, ability to form films, chelate metal ions and specific structural characteristics[4]. CS, a linear randomly distributed nitrogeneous polysaccharide has being reported to possess either a rigid rod-type structure[5,6] or a semi-flexible-coil[7-10]. It has also been indicated that the flexibility associated with CS is moderately influenced by degree of acetylation[8].

The degree of deacetylation (DD) of molecular chain of chitin can be increased by increasing the temperature or strength of the alkaline solution during its deacetylation. The DD can also be determined by its ratio of N-acetylglucosamine to N-glucosamine structural units. When the number of N-acetylglucosamine units is more than 50%, the biopolymer is said to be chitosan and when the number of N-acetylglucosamine units are higher, the polymer is said to be chitin.

The solubility of CS can be modified by changing the DA or by modifying the pH and ionic strength of the formulation. In neutral pH, CS molecules losses their charge and get precipitated from the solution. The properties of CS are greatly affected by the conditions under which it is processed, because it is the process conditions that control the amount of deacetylation that occurs. The DD controls the amount of free amino groups in the polymer chain. The free amino groups give CS its positive charge in acidic environment. The amino group and the hydroxyl group give CS its functionality which allows it to be a highly reactive glycan. CS positive charge allows it to have many electrostatic interactions with negatively charged molecules.

The uniqueness of chitosan is that, it is a highly basic glycan due to the presence of unbounded (lone) pair of electrons on the primary amine nitrogen. In contrast, most naturally occurring glycans (e.g. carragenans, agarose, agar, alginic acid, pectin, dextran and cellulose) are neutral or acidic in nature. This singular difference, in addition to the presence of primary hydroxyl group and secondary hydroxyl group (also available in other glucans) stands out CS from the other polysaccharides. The presence in the repeat unit of a $-NH_2$ in a C-2 position of CS is especially interesting for its controlled chemical modification. Another significance of CS is its water solubility in acidic conditions ($pH < 6$) thereby allowing the preparation of its derivatives with optimized biocompatible and biodegradable properties in homogeneous conditions. CS is the only polycationic polymer and its derivatives have received a great deal of attention in the food, cosmetic and pharmaceutical industries. Other applications include waste-water treatment, antibacterial, antitumor and anticoagulant properties[11]. In an aqueous acidic medium, CS is instantly solubilized due to the removal of the acetyl moieties present in the amine functional groups. In comparison, solubility is more pronounced in organic acids when compared to mineral (inorganic) acids. Solubilization occurs as a consequence of the protonation of $-NH_2$ fuunctional groups on the C-2 position of D-glucosamine residues[12].

5.3 Chemically Modified Chitosan for Drug Delivery Systems

CS has been regarded as a source of potential bioactive material, but it also has several limitations to be utilized in biological system, including its poor solubility under physiological condition. In order to overcome these limitations, chemical modifications of CS structures have resulted in increased solubility in water as well as in organic solvents have been reported by some investigators. Chemical modification of various reactive groups (amino, hydroxyl) on CS provides a powerful means to promote new biological activities and to modify its mechanical properties.

The primary amino groups on the molecule are reactive and provide a side for group attachment using a variety of mild reaction conditions. The effect of addition of a side chain is to disrupt the crystal structure of the material and hence increase its amorphousness. The consequence is that it lowers the stiffness of the material (CS) and altered its solubility. The exact nature of the changes in chemical and biological properties depends on the nature of the side group. Additionally, the characteristic features of CS such as being cationic, insoluble at high pH, can be completely altered by a sulphation process which can render the molecule anionic and water soluble, and also introduce anti-coagulant properties[13]. Almost unlimited number of functional groups could be attached to CS, and these side groups can be chosen to provide specific functionality and modify physical properties or altered physicochemical properties.

The solubility of chitosan can be decreased by cross-linking the polymer with covalent bonds like glutaraldehyde. Covalent cross-linking leads to formation of hydrogels with a permanent network structure, since irreversible cross-linking are formed. This type of linking allows absorption of water and/or bioactive compounds without dissolution and permits drug release by diffusion. pH-controlled drug delivery is made possible by the addition of another polymer. Ionically cross-linked hydrogels are generally considered as biocompatible and well-tolerated. Their non-permanent network is formed by reversible links. Ionically cross-linked CS hydrogels exhibit a higher swelling sensitivity to pH changes when compared to covalently cross-linked CS hydrogels. This extends their potential applications, since dissolution can occur in extreme acidic or basic pH conditions[14]. The swelling power of the CS decreases with an increase in the concentration of the cross-linking agent[15]. The swelling capacity of enzyme-containing films decreases more pronouncedly, probably because of formation of additional crosslinks due to the participation of the functional groups of the enzyme in the reaction. The cross-linking of CS chains by glutaraldehyde was also characterizes by coloration and decreased aqueous solubility.

The nitrogen in CS is mostly in the form of primary aliphatic amino group. It therefore undergoes reactions typical of amines such as N-acylation and Schiff reaction. The amino groups are introduced into CS nitrogen by N-acylation with acyl halides or acid anhydrides. Fully acetylated chitin can be obtain by reacting CS with acetic anhydride [$(CH_3CO)_2O$]. Linear aliphatic N-acyl groups higher than propionyl permit rapid acetylation of the hydroxyl groups in CS.

CS reacts with aldehydes and ketones forming aldimines and ketimines respectively, at ambient temperature. Reaction with ketoacids followed by reduction with sodium borohydride ($NaBH_4$) produces glucans carrying proteic and non-proteic amino acid groups. Examples of non-proteic amino acid glucans obtained from CS are the N-carboxybenzyl chitosans derived from O- and P-phthalaldehydic acids. N-Carboxymethyl CS is derived from glyoxylic acid. CS and simple aldehydes produce N-alkyl CS upon hydrogenation. The presence of more or less bulky substituent weakens the hydrogen bonds of CS; therefore N-alkyl CS swell in water inspite of the hydrophobicity of the alkyl chains, but they retain the film forming property of CS[16]. The amino group at the C-2 position of CS makes it more versatile than chitin.

CS can be readily derivatized by utilizing the reactivity of the primary amino group and the primary and secondary hydroxyl groups. It has a large number of applications in drug delivery systems. Its applicability can be further exploited by modification of basic structure to obtain modified polymers with a wider range of properties. Some of the useful modified CS are listed as follows:

(i) ***Chitosan conjugates***: CS can be reacted with bioactive excipients for delivery of active ingredients such as calcitonin. Chitosan-4-thiobutylamidine conjugate and 5-methylpyrrolidinone CS are CS conjugate processing mucoadhesive properties. In an experiment conducted by Guggi and Bernkop[17], they attached enzyme inhibitor to CS. The polymer obtained, retained its mucoadhesive properties and the attached enzme inhibitor prevent drug degradation by inhibiting enzymes (e.g., chymotrysin and trypsin). The delivery of sensitive peptide drugs (e.g. calcitonin) was enhanced by this conjugated CS.

(ii) ***Chitosan esters***: Reaction of CS with glutamic acid, succinic anhydride and phthalic anhydride result in the formation of CS (glutamate, succinate, and phthalate) respectively. These esteric forms of CS have different degree of solubility and are insoluble in acidic environment. In contrast, these esters provide sustained release in basic conditions[18].

(iii) ***N-Trimethylene chloride chitosan***: N-Trimethylene chitosan chloride (TMC) is a quaternary derivative of CS and it has a superior aqueous solubility, intestinal permeability as well as higher absorption of neutral and cationic peptide analogue over a wide pH range. The TMC polymer is stipulated according to their degree of methylation such as TMC -20%, TMC-40% and TMC-60%. Decrease in solubility has been observed in TMC with higher degrees of substitution. According to Thanou et al.[19] quaternization of TMC decreases the transepithelial electrical resistant and thereby influences its drug absorption-enhancing properties.

(iv) ***Lactic/glycolic acid chitosan hydrogels***: The synthesis of CS hydrogels in the absence of catalyst was carried out by direct grafting of D, L-lactic and /or glycolic acid onto CS[20]. These researchers established that a stronger interaction existed between water and CS chains after grafting with lactic acid and /or glycolic acid. The side chain could aggregate and form physical cross-linking which result in pH-sensitive CS hydrogels. This indicate that at high pH chitosan's amines are reactive allowing a range of chemistries to be employed to graft substituents to functionalize CS. These CS hydrogels are considered to be potentially useful for biomedical applications such as wound dressing and drug delivery systems, since both polyester side chains and CS are biocompatible and biodegradable[21].

(v) ***Dendronized chitosan-sialic acid hybrids***: Sashiwa et al.[22] synthesized dendronized chitosan-sialic acid hybrids by using gallic acid and tri (ethylene glycol) in order to improve the water solubility of CS. The water solubility of these unique derivatives were further improved by N-succinylation of the remaining amine functionality.

(vi) ***N-phthaloylation of chitosan***: Due to the poor solubility of CS in some organic solvents, it is chemically modify by N-phthaloylation. This process is effective for solubilization since it attaches a bulky group to the rigid backbone and breaks hydrogen atoms on the amine groups to prevent hydrogen bonding. Fully deacetylated chitosan was treated with phthalic anhydride in DMF to give N-phthaloyl-chitosan. It was readily soluble in polar organic solvents.

(vii) ***Cds quantum dots (QDs) chitosan biocomposite:*** CS derivatives of Cds QDs improve aqueous solubility and stability of chitosan. They also enhance the thermal decomposition of CS. An effective synthesis of Cds QDs CS biocomposite was carried out by mixing CS with Cd $(Ac)_2$ and subsequently dissolved in 1 percent HAc aqueous solution, followed by the treatment with Cds and thus smooth, flat, yellow Cds QDs CS composite films were obtained.

5.4 Importance of Mucoadhesive Properties of Chitosan in Drug Delivery Systems

Mucoadhesion is the specific term for adhesion when one of the surfaces is mucus. Mucus consists largely of water (> 90%) and the high molecular weight glucoprotein, mucin[23,24]. The key sugar residues for mucoadhesive interaction are the acidic ones (sialic acid, and some sulphated galactose) and the hydrophobic methyl containing fucose. CS interacts strongly with the negatively charged sialic acid residues[25] although hydrogen bonding and hydrophobic interactions are also significant[26]. This latter mechanism involves the interaction between positively charged amino groups and the negatively charged mucus gel layer. Many factors such as physiological variables and physicochemical properties of CS alter its mucoadhesion. CS is positively charged in the stomach due to the acidic medium and interacts with the negatively charged mucin by electrostatic forces. The extent of this bonding depends on the amount of sialic acid in the mucin and the DD of CS.

Since the interaction of CS with mucin is higher at acidic and slightly acidic pH because it is at this pH that the CS charge is positive, pH is an important factor in determining the strength of the reaction. It was also reported that CS with higher molecular weight (MW) penetrates more into the mucin layer, indicating that mucoadhesion is stronger the higher the MW in the case of CS. The DD and MW are therefore important properties for mucoadhesion. Mucoadhesive properties were also enhanced when CS cationic derivatives such as N-trimethyl CS chloride and cyclodextrin-CS complexes were utilized in experiment[27,28]. In contrast CS was shown not to be affected by the environment of the small intestine[29] perhaps due to the high pH (alkaline) medium.

5.5 Usage of Chitosan and Modified Chitosans in Drug Delivery Systems

The cationic, biodegradability, non-toxicity, biocomptability, antibacterial, antifungistic, and bioadhesive properties of CS makes it suitable to deliver drugs to their appropriate location within a biological system. The ability of CS to degrade within a biological system is based on the degree of de-acetylation that occurs in it during processing. This allows drugs to be released into the body in a controlled manner and as effective as possible. The free amine group that gives CS its positivity in acidic environment is significant to drug delivery for it is this cation that permits it to interact with negatively charged (anion) polymers, drugs and other organic bioactive molecules. It is the property responsible for chitosan's mucoadhesion and especially important for drug delivery. Its attractiveness for usage in drug transport is due to its capability to be used in different forms, e.g., as gels, colloidal particles, and copolymer. Chitosan's versatility along with its other physicochemical properties including its bioadhesivity begets a biomaterial well suited for drug delivery[30].

(i) ***Vaginal drug delivery***: The vagina is a potential site for drug delivery due to its rich blood supply and large surface area[31] and well understood microflora[32] (Valenta, 2005). The introduction of thiol groups into the primary amine groups of CS increases its mucoadhensive properties. The thiol modified CS, embeds clotrimazole and it is widely used in the treatment of mycotic vaginal infections[33]. Vaginal tablets of CS containing metronidazole and acriflavine have showed adequate release and good adhesion properties[34].

Drug delivery release rates may vary during the menstrual cycle and this is especially important at the menopause[32]. The drug delivery systems are based on mucoadhesion and the vaginal route has been indicated to be favorable in the delivery of many drugs, e.g., propranolol and human growth hormone[32]. Despite the fact that the vaginal route offers many advantages, its main disadvantages being that it is only available to females.

(ii) ***Oral drug delivery***: The route from the delivery of drug that is still the most popular with patient is through the mouth and down the alimentary canal tract. The oral route is of particular interest as it results in less pain, greater convenience, higher compliance and reduced infection risk, when compared to subcutaneous injections[35]. An ideal buccal delivery system should stay in the oral cavity for few hours and release the drug in a unidirectional way toward the mucosa in a controlled or sustained-release fashion. Mucoadhensive polymers prolong the residence time of the device in the oral cavity[36], while bilayered devices ensure the release of the drug occurs in a unidirectional way. Buccal patches, tablets, and gel formulations prepared with CS have effectively delivered the drug unidirectionally into systemic circulation through buccal mucosa. CS is an excellent polysaccharide to be used for buccal delivery due to its muco/bioadhesive properties and can act as an absorption enhancer[17]. It has also been reported that chitosan microparticles with no drug incorporated possess anti-microbial activities. The promising unique mucoadhesive and absorption enhancing quality of this polymer further confirmed its aptness for the buccal drug delivery.

However, limitations associated with the oral route of administration include low bioavailability due to relativity low passage of active agents across the mucosal epithelium, rapid polypeptide degradation due to enzymic digestion in the gastrointestinal tract, enzymatic proteolysis and acidic degradation of orally administered drug in the stomach[37]. Different methods have been deviced to increased local penetration by using permeation enhancers[38], protease inhibitor and enteric coatings from enzymatic proteolysis and acidic degradation in the colon which has resulted in a concentrated effort to target their delivery to this organ[18,39].

(iii) ***Nasal drug delivery***: The clearance time through the whole alimentary tract is generally to short, rendering oral drug administration a very ineffective process, with much of the drug unabsorbed. More recently interest has focused on drug absorption through nasal epithelia, which results in very rapid absorption. CS, a cationic bioadhesive natural polymer as a remarkable ability to increase the transport of polar drugs, peptides and proteins across epithelial surfaces. Various CS salts, such as CS lactate, CS as aspartate, CS glutamate, and CS hydrochloride are good modified CS for nasal sustained release of vancomycin[40]. Research showed that bioadhesive CS microspheres of pentezocine for intranasal systemic delivery significantly improved the bioavailability with sustained and controlled blood level profiles compared to intravenous and oral administration[41]. CS delivery systems have two central effects on nasal mucose thereby promoting drug permeation. The first is the clearance of the formulation from the nasal cavity is abridged by the presence of cation in the CS that bind to nagetively charged sialic residues that renders excellent mucoadhesive properties and consequently prolong contact time. The second is the reversible and momentary action on epithelial tight junctions between cells that paracellularly steps the drug transportation. Shaoyun et al.[42] has also reported nasal insulin delivery in chitosan solution.

(iv) ***Gene delivery***: Due to its positive charge (cation), CS has the tandency to interact with negative molecules such as deoxyribonucleic acid (DNA). CS can effectively bind DNA and protect it from nuclease degradation. The utilization of CS as non-viral vector for gene delivery offers many merits compared to viral vectors. Additionally, CS does not produce endogenous recombination, oncogenic effects or immunological reactions[43]. Another advantage is the cheap production of CS/DNA complexes. DNA loaded CS microparticles were found to be stable during storage. The application of DNA/CS nanospheres has advanced *in vitro* DNA transfection research and the data has showed their importance for gene delivery[44].

Conclusion

Chitosan is a natural copolymer of glucosamine and N-acetylglucosamine and it has been described as a non-toxic, biodegradable and biocompatible polymer with very interesting biological properties, such as permeation-enhancing and mucoadhesive properties, anticoagulate and antimicrobial activity. These unique properties make chitosan an excellent material for the development of new industrial applications. CS is a versatile polymer and due to the present of reactive groups (NH_2, OH) on its backbone, it can be easily modified to improve its properties or to open the use of CS in new applications.

The development of new delivery system for the control release of drug is one of the most interesting fields of research in pharmaceutical sciences. There are several unique fundamental properties of chitosan that are useful in solving drug delivery problems, as it can be combined with the drug covalently or ionically to overcome problems like solubility, stability or permeability.

Many patients benefit from advanced drug delivery systems, receiving safer and more effective doses of the medicines, they need to fight a variety of human ailments. Controlled drug delivery occurs when a polymer like chitosan is judiciously combined with a drug or other active agent in such a way that the active agent is release from the material in a predesigned manner. The release of active agent may be constant over a long period, it maybe cyclic over a long period or it may be trigger by the environment or other external events. In any case, the purpose behind controlling the drug delivery is to achieve more effective therapies while eliminating the potential for both under and over dosing.

In the last decade, there has been a great deal of interest in the use of polysaccharides like chitosan in drug delivery systems. CS show great potential, however, many important issues still remain unsolved. These include: the stability of chitosan, with the important constraint that, chitosan are soluble only at pH less than 6; their construction into microparticles capable of surviving the large environment variation between mouth and intestine for oral drug delivery. In addressing these issues, the optimal degree of acetylation and MW of the CS need to be addressed.

Further studies need to include the stability of drugs and vaccines in chitosan formulations, and stability of the complex in the presence of gut bacteria; the optimal epithelial site of uptake for a given purpose. The optimal particle size for uptake by the desired cell type; and the requirement for bolus versus sustained delivery of mucosal vaccines.

References

1. Bansal, V., Sharma, P. K., Sharma, N., Pal, O. P., & Malviya, R. (2011). Applications of chitosan and chitosan derivatives in drug delivery. Advances in Biological Research, 5 (1), 28-37.
2. Malviya, R., Srivastava, P., Bansal, V., & Sharma, P. K. (2010a). Formulation, evaluation and comparison of sustained release matrix tablets of diclofenac sodium using natural polymers as release modifier. International Journal of Pharmaceutical and Biological Science, 1 (2), 1-8.
3. Malviya, R., Srivastava, P., Bansal, V., & Sharma, P. K. (2010b). Preparation and evaluation of disintegrating properties of *Cucurbita*

maxima pulp powder. International Journal of Pharmaceutical Science, 2 (1), 395-399.

4. Ravi Kumar, M. N. V. (2000). A review of chitin and chitosan applications. Reactive and Functional polysaccharides, 46, 1-27.
5. Morris, G. A., Castile, J., Smith, A., Adams, G. G., & Harding, S. E. (2009). Macromolecular conformation of chitosan in dilute solution: a new global hydrodynamic approach. Carbohydrate Polymers, 76, 616-621.
6. Kasaai, M. R. (2006). Calculation of Mark-Houwink-Sakurada (MHS) equation viscometric constants for chitosan in any solvent-temperature system using experimental reported viscometric constants data. Carbohydrate Polymers, 68, 477-488.
7. Schatz, S., Viton, C., Delair, T., Pichot, C., & Domard, A. (2003). Typical physicochemical behaviors of chitosan in aqueous solution. Biomacromolecules, 4, 641-648.
8. Mazeau, K., & Rinaudo, M. (2004). The prediction of the characteristics of some polysaccharides from molecular modeling. Comparison with effective behavior. Food Hydrocolloids, 18, 885-898.
9. Lamarque, G., Lucas, J.-M., Viton, C., & Domard, A. (2005). Physicochemical behavior of homogeneous series of acetylated chitosans in aqueous solution: role of various structural parameters. Biomacromolecules, 6, 131-142.
10. Velasquez, C. L., Albornoz, J. S., & Barrios, E. M. (2008). Viscosimetric studies of chitosan nitrate and chitosan chlorhydrate in acid free NaCl aqueous solution. E-Polymer, 014.
11. Muzzarelli, R. A. A. (2009). Genipin-crosslinked chitosan hydrogels as biomedical and pharmaceutical aids. Carbohhydrate Polymers, 77, 1-9.
12. Morris, G. A., Kok, M. S., Harding, S. E., & Adam, G. G. (2010). Polysaccharide drug delivery systems based on pectin and chitosan. Biotechnology and Genetic Engineering Reviews, 27, 257-284.
13. Suh, J. K. F., & Mathew, H. W. T. (2000). Application of chitosan-based polysaccharide biomaterials in cartilage tissue engineering: a review. Biomaterials, 21, 2589-2598.
14. Berger, J., Reist, M., Mayer, J. M., Felt, O., Peppas, N. A., & Gurny, R. (2004). Structure and interactions in covalently and ionically cross-linked chitosan hydrogels for biomedical applications. European journal of Pharmaceutics and Biopharmacutics, 57, 19-34.
15. Dutta, P. K., Dutta, J., & Tripathi, V. S. (2004). Chitin and chitosan: Chemistry, properties and application. Journal of Scientific and Industrial Research, 63, 20-31.

16. Dutta, P. K., Ravikumar, M. N. V., & Dutta, J. (2002). Chitin and chitosan for versatile applications. J. M. S. Polymer Rev., 42, 307.
17. Guggi, D., & Bernkop-Schnurch, A. (2003). *In vitro* evaluation of polymeric excipients protecting calcitonin against degradation by intestinal serine proteases. International Journal of Pharmaceutics, 252 (1-2), 187-196.
18. Sinha, V. R., & Kumria, R. (2001). Polysaccharides in colon-specific drug delivery. International Journal of Pharmaceutics, 224 (1-2), 19-38.
19. Thanou, M., Verhoef, J. C., Marbach, P., & Junginger, H. E. (2000). Intestinal absorption of octreotide: N-trimethyl chitosan chloride (TMC) ameliorate the permeability and absorption properties of the somatostation analogue *in vitro* and *in vivo*. Journal of Pharmaceutical Science, 89 (7), 951-957.
20. Qu, X., Wirsen, A., & Albertsson, A. C. (2001). Effect of lactic/glycolic acid side chains on the thermal degradation kinetics of chitosan derivatives. Polymer, 41, 4841.
21. Mitsumata, T., Suemitcu, Y., Fuji, K., Taniguchi, T., & Koyama, K. (2003). pH-response of chitosan, k-carrageenan, carboxymethyl cellulose sodium salt complex hydrogels. Polymer, 44, 7103.
22. Sashiwa, H., Shigemasa, Y., & Roy, R. (2001). Chemical modification of chitosan: synthesis of dendronized chitosan-sialic acid hydrid using convergent grafting of preassembled dendrons built on gallic acid and tri (ethylene glycol) backbone. Macromolecules, 34, 3905.
23. Harding, S. E. (2003). Mucoadhesive interactions. Biochemical Society Transactions, 31, 1036-1041.
24. Harding, S. E. (2006). Trends in mucoadhesive analysis. Trends in Food Science and Technology, 17, 255-262.
25. Dodou, D., Breedveld, P., & Wieringa, P. A. (2005). Mucoadhesive in gastrointestinal tract: revisiting the literature for novel applications. European Journal of Pharmaceutics and Biopharmaceutics, 60, 1-16.
26. Sogias, I. A., William, A. C., & Khutoryansky, V. V. (2008). Why is chitosan mucoadhesive? Biomacromolecules, 9, 1837-1842.
27. Van der Merwe, S. M., Verhoef, J. C., Verheijden, J. H. M., Kotze, A. F., & Junginger, H. E. (2004). Trimethylated chitosan as polymeric absorption enhancer for improved peroral delivery of peptide drugs. European Journal of Pharmaceutics and Biopharmaceutics, 58, 225.
28. Venter, J. P., Kotze, A. F., Auzely-Velty, R., & Rinaudo, M. (2006). Synthesis and evaluation of the mucoadhesivity of a CD-chitosan derivative. International Journal of Pharmaceutics, 313, 36.

29. Okamoto, Y., Nose, M., Miyatake, K., Sekine, J., Oura, R., Shigemasa, Y., Minami, S. (2001). Physical changes in chitin and chitosan in canine gastrointestinal tract. Carbohydrate Polymers, 44, 211-215.
30. Prabaharan, M., Reis, R. L., & Mano, J. F. (2005). Chitosan-based particles as controlled drug delivery systems. Drug Delivery, 12 (1), 41-57.
31. Vermani, K., & Garg, S. (2000). The scope and potential of vaginal drug delivery. Pharmaceutical Science & Technology Today, 3, 359-364.
32. Valenta, C. (2005). The use of mucoadhesive polymers in vaginal delivery. Advanced Drug Delivery Reviews, 57, 1692-1912.
33. Kast, C. E., & Bernkop-Schnurch, A. (2001). Thiolated polymer-thimers: Development and *in vitro* evaluation of chitosan-thiglycolic acid conjugates. Biomaterials, 22, 2345-2352.
34. Sakkinen, M., Marvola, J., Kanerva, H., Lindevall, K., Ahonen, A., & Marvola, M. (2004). Scintigraphic verification of adherence of a chitosan formulation to the human oesophagus. European of Pharmaceutics and Biopharmaceutics, 57, 145-147.
35. Yadav, N., Morris, G. A., Harding, S. E., Ang, S., & Adams, G. G. (2009). Various non-injectable delivery systems for the treatment of diabetes mellitus. Endocrine, Metabolic and Immune Disorders-Drug Targets, 9, 1-13.
36. Aiedeh, K., & Taha, M. O. (2001). Synthesis of iron-cross-linked hydroxamated chitosan succinate and their in-vitro evaluation as potential matrix materials for oral theophylline sustained-release beads. European Journal of Pharmaceutical Science, 13(2), 159-168.
37. Lin, Y. H., Chen, C. H., Liang, H. F., Kulkarni, A. R., Lee, P. W., Chen, C. H., & Sung, H. W. (2007). Novel nanoparticles for oral insulin delivery via the paracellular pathway. Nanotechnology, 18, 1-11.
38. Carino, G. P., Jacob, J. S. & Mathiowitz, E. (2000). Nanosphere based oral insulin delivery. Journal of Conrolled Release, 65, 261-269.
39. Chambin, O., Dupuis, G., Champion, D., Voilley, A., & Pourcelot, Y. (2006). Colon-specific drug delivery: Influence of solution reticulation properties upon pectin beads performance. International Journal of Pharmaceutics, 321, 86-93.
40. Thanou, M., Henderson, S., Kydonieus, A., & Elson, C. (2007). N-sulfonato, N, O-carboxymethylchitosan: a novel polymeric absorption enhances for the oral delivery of macromolecules. Journal of Controlled Release, 117 (2) 171-178.
41. Kean, T., Roth, S., & Thanou, M. (2005). Trimethylated chitosans as non-viral gene delivery vectors: cytotoxicity and transfection efficiency. Journal of Controlled Release, 103 (3), 643-653.

42. Shaoyun, Y., Ying, Z., Fenglan, W., Xuan, Z., Wanliang, L., Haa, Z., & Qiang, Z. (2004). Nasal insulin delivery in the chitosan solution: *in vitro* and *in vivo* studies. International Journal of Pharmaceutics, 281, 11-23.

43. Ferber, D. (2001). Gene therapy: safer and virus-free? Science, 294, 1638-1642.

44. Corsi, K., Chellat, F., Yahia, L., & Fernandes, J. C. (2003). Mesenchymal stem cells, MG63 and HEK293 transfection using chitosan/DNA nanoparticles. Biomaterials, 24, 1255-1264.

6 Alginate, Its Chemistry and Applications in Advanced Drug Delivery

Sukrutha Suresh, Gurmeet Singh, Kathyayini Hanumanthaiah, Krishna Venkatesh and Ravi Kumar Kadeppagari*

Nanobiosciences, Centre for Emerging Technologies, Jain University, Jain Global Campus, Jakkasandra, Kanakapura Main Road, Ramanagara Dist., Karnataka, 562112.

6.1 Introduction

Polysaccharides are the polymers of monosaccharides which are linked by glycosidic bonds. Polysaccharides possess various favorable characteristics with respect to pharmaceutical applications such as biocompatibility, low toxicity, stability, hydrophilic nature and availability of reactive sites for the chemical modification. Polysaccharides are produced from plants (cellulose, pectin and guar gum), algae (alginate and carrageenan), microbes (dextran and xanthan gum) and animals (Chitosan, hyaluronan, chondroitin and heparin). Alginate, a polysaccharide that consist of 1,4-linked β-D-mannuronic acid & α-L-guluronic acid moieties has the ideal characteristics for the drug delivery like inert, strong, biocompatible, easy to fabricate & sterilize, comfortable to administer and capable of loading high levels of desired drug. Alginate is the term usually used for the salts of alginic acid, but it can also be referred to the derivatives of alginic acid.

The first scientific studies on the extraction of alginate from brown seaweed were made by the British chemist E.C. Stanford at the end of the 19th century and the large-scale production of alginate was introduced 50 years later. Alginate is widely used in the food, pharmaceutical, cosmetic and textile industries. Alginate matrices have the ability to deliver the bioactive molecules inside the cells without losing their biological activities[1,2]. Many methods have been implemented for the encapsulation of alginate with nanoparticles that used for the delivering a

* *Corresponding Author*

drug to a particular organ or a tissue. This review will cover the sources, commercial production, chemistry, properties, structure and the applications of alginates in the drug delivery.

6.2 Sources of Alginate

Alginate is the most abundant biopolymer obtained from brown seaweeds and major proportion is found in the cell walls and intracellular matrix. Alginate is found in the intracellular matrix of the kelp species and in the cell wall of brown algae (*Phaeophyceae*). it constitutes up to 40% of the dry weight in both cases. The alginate forms are mixed with various cations naturally found in sea water including Mg^{+2}, Ca^{+2}, Sr^{+2}, Ba^{+2} and Na^{+}. The native species is usually found as an insoluble Ca^{+2} cross-linked gel[3].

Alginates are commercially produced from *Laminaria digitata, Macrocystis pyrifera* and *Ascophyllum nodosum.* Scotland, Ireland, Norway, France, China, Japan and Korea are the main countries that harvest *Laminaria digitata.* It is very popular as a food in Japan and only the material that is unsuitable for food is used for alginate extraction. The cost of cultivated *Laminaria* is higher than the harvested wild material. However Chinese are able to absorb these higher costs and use the cultivated product for alginate extraction. *Macrocystis pyrifera* is harvested on the west coast of North America. *Ascophyllum nodosum* grows in the intertidal zone and it has been harvested by hand in Scotland and Ireland for more than a century. It is also harvested in the southern parts of Nova Scotia[4].

6.3 Commercial Production of Alginate

The commercial production of alginate from algae involves three steps namely, removal of counter ions, solubilization and precipitation. In the first step, the counter ions will be removed by using 0.1-0.2 M HCl. In the second step, alginate will be solubilized by neutralizing with either sodium carbonate or sodium hydroxide. Then, the sodium alginate is precipitated by using alcohol, $CaCl_2$ or HCl. Alginate will be milled after milling. Alginates obtained after this process contain mitogenic and cytotoxic impurities limiting their use in the biomedical applications. These impurities will be removed by either free flow electrophoresis[5] or chemical extraction method[6].

6.4 Chemistry of Alginate

Alginate is a linear un-branched polysaccharide that consists of varying amounts of 1, 4-linked β-D mannuronic acid and α-L guluronic acid residues and the residues are arranged like blocks along the chain. A polymer contains three kinds of blocks namely, G- block which consists of L- guluronic acid, M- block that contains D- mannuronic acid and MG- block which consists of both the residues[7,8,9]. The physical properties of alginate depend on the relative proportion of these blocks and this composition depends on the organism from which alginate is isolated. For example, alginate that is isolated from *L.hyperboea* kelp has high α-L guluronic acid residues, whereas alginate isolated from *A.nodosum* and *L.japonica* has low content of α-L-guluronic residues. The composition determines the molecular weight and the physical properties of the polymer. Chemical structures of different blocks of alginate are represented in the Fig. 6.1.

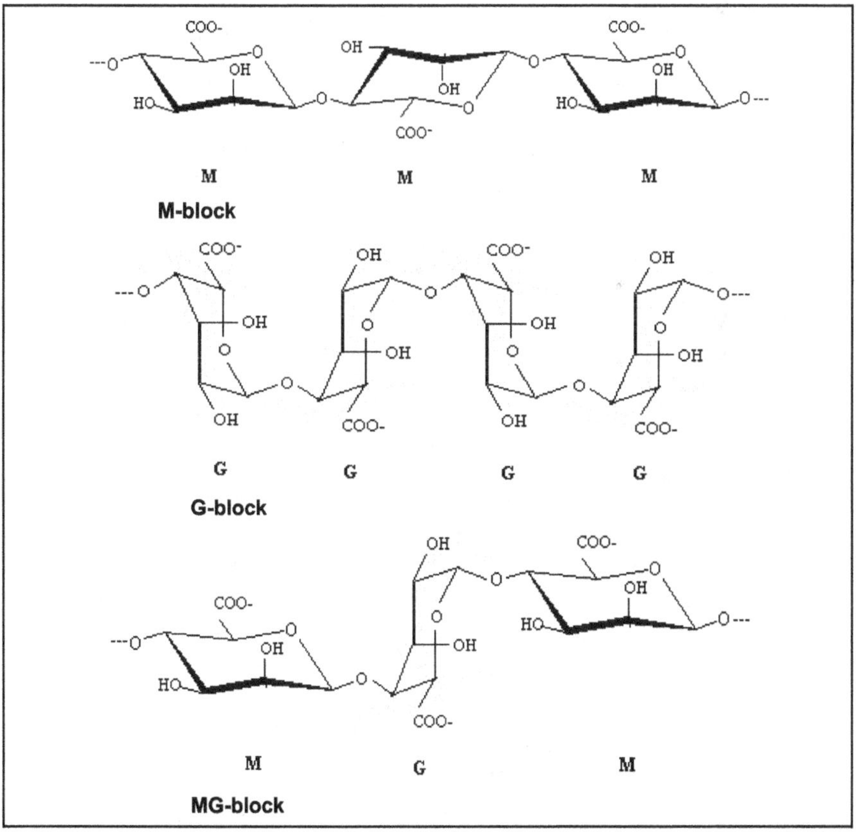

FIGURE 6.1 Chemical structures of different alginate blocks.

6.5 Physical Properties of Alginate

1. ***Solubility***: Alginate solubility depends on the three parameters namely, solvent pH, ionic strength and gelling ions. Alginates are soluble above a certain critical pH value and in order to be alginates soluble their COOH groups have to be deprotonated. Changes in the ionic strength of the medium affect the soluble properties like polymer conformation, chain extension and viscosity[10]. The presence of divalent cations such as Ca^{+2}, Sr^{+2} and Ba^{+2} will lead to the gelling of alginates. Hence, aqueous solvent has to be free from cross-linking ions in order to dissolve the alginates.
2. ***Molecular weight***: Commercially available sodium alginate molecular weights range from 32,000 to 400,000 g/mol. The increase in the molecular weight of the alginate will improve the physical properties of the gel. But, the alginate that is formed from the high molecular weight polymer becomes highly viscous which is not suitable for processing and biomedical applications[11,12].
3. ***Viscosity***: Over the pH range of 5-11 viscosity of the alginate solutions is not affected. The viscosity of alginate solutions increases with decrease in the pH and reach maximum around 3.0-3.5 since at low pH the free COO^- ions get protonated and form COOH and due to this there will be reduction in the electrostatic repulsion between the chains. This will enable the chains to come closer, form hydrogen bonds and produce high viscosity. When the pH is reduced to 3- 4 gels are formed. When the pH is above 11, depolymerization occurs slowly and viscosity of the alginate solution decreases[13].

6.6 Chemical Properties of Alginate

Alginate undergoes hydrolytic cleavage under acidic conditions and the glycosidic bonds are cleaved[14]. Alginates will undergo degradation not only at acidic or basic condition, but also at neutral pH in the presence of reducing agents[10]. Alginate of different species of brown algae contains different amount of phenolic compounds and the degradation rate of alginate is higher in case of the species that contains higher phenolic content[15,16]. Alginate undergoes cleavage to form unsaturated compounds in the presence of lyase enzyme[17,18]. Reducing compounds like sodium hydrogen sulfide, ascorbic acid, cysteine caused the degradation of alginates by forming peroxide and free radicals[19]. Alginate degradation also occurs when they are subjected to sterilization methods like autoclaving, heat treatment, exposing to γ- irradiation and ethylene dioxide treatment[20]. Alginates can be modified using two functional groups, OH (at C-2 and C-3 positions) and COOH (at C-6 position). Different alginates can be derived by various modifications like, acetylation, phosphorylation, sulfation, hydrophobic modification, attaching of ligands and graft polymerization.

6.7 Biocompatibilty

Alginates that contain high M content are immunogenic and about ten times more potent than G alginates in inducing cytokine production[21]. The immunogenic response to the alginates was attributed to the impurities that present in the alginates. Alginates have the impurities like heavy metals, phenolic compounds, toxins and proteins. There was no significant inflammatory reaction when highly purified alginate was used[22,23]. Hence, highly purified alginates have to be used for the medical applications. Alginate polymer has a very strong bioadhesive property that helps in the mucosal delivery of drugs[24].

6.8 Gelling Methods

Alginates are used in the form of hydrogels for biomedical applications like, drug delivery, tissue engineering and wound healing. Hydrogels are 3-D cross-linked matrices consisting of water rich hydrophilic polymers. Hydrogels are biocompatible and they are typically obtained by cross-linking the hydrophilic polymers through chemical or physical means. Some of the methods are described below.

1. ***Ionic cross-linking***: In this method, hydrogels prepared from aqueous alginate solution by using the divalent cations like Ca^{+2}, Sr^{+2} and Ba^{+2}. The gel is formed by the interaction between the guluronate blocks and the divalent ions and these guluronate blocks form junction between the guluronate block of one polymer and that of next polymer chain. This cross-linking is termed as egg-box model cross-linking[25]. Calcium ions are the most commonly used divalent ions to form an alginate gel. The cross-linking of calcium can be done by diffusion method or internal setting. The drawback with these gels is limited stability under physiological conditions due to the release of divalent cations into the surrounding media in exchange with monovalent cations.

2. ***Covalent cross-linking***: This cross-linking is done in order to improve the physical properties of the alginate gel for various applications like tissue engineering. Covalent cross linking of alginate leads to the formation of alginate matrices with enhanced mechanical properties. Covalent cross-linking will lead to more stable gels and due to this it is preferable over ionic cross-linking. Mechanical properties and swelling of alginate gels can be strictly controlled by using different cross-linking molecules like ploy ethylene glycol (PEG), poly acrylamide-co-hydrazide (PAH) and adipic acid dihydrazide (AAD). However, the cross-linking molecules could be toxic and free molecules need to be thoroughly removed from the gels before using for the biomedical applications[13].
3. ***Thermal gelation***: Hydrogels that are sensitive to temperature changes have been explored for drug delivery due to their on-demand adjustament for the release of drug. Graft copolymerization of N-isopropylacrylamide onto the alginate after reacting with ceric ions results into the temperature-sensitive alginate gels that show sensitivity near body temperatura[13].
4. ***Cell cross-linking***: Here cells will cross link the alginate chains that modified with cell adhesive ligands. RGD modified alginate chains form a uniform, reversible network in the presence of cells without requiring any extra cross-linking molecules[26,27]. This could be an ideal cell delivery system during tissue engineering.

6.9 Applications of Alginate in the Advanced Drug Delivery

Alginate has application in the preparation of oral dosage forms like tablets and capsules and it has been mainly applied in the diffusion based delivery systems. Alginate based diffusion systems are divided into two main types namely polymer membrane and polymer matrix systems. In the polymer membrane system, the drug will be encapsulated within the drug reservoir compartment and in the polymer matrix system the drug is dispersed homogenously in a rate controlling polymer matrix. Alginates are also used in the preparations for local administration of drugs like wound dressings. In yet another application they can be used as drug excipients[28]. Here we describe the advances in the controlled drug delivery using alginates and their derivatives.

Alginates are widely used for the delivery of small chemical drugs or bioactive molecules. Alginate gels usually have the pores of around 5 nm size[29] due to which small molecules will be easily diffused out of the gels. Flurbiprofen will be completely released from the ion cross-linked and partially oxidized alginate gels within one and half hour[13]. Alginate gels can also be used for the sequential or simultaneous delivery of the multiple drugs. Anti-cancer drug, methotrexate that is non-interactive with alginate will quickly diffuse out of the gel; while another anti-cancer drug, doxorubicin that is attached covalently to the alginate will be released after the hydrolysis of the cross-linker and yet another drug, mitoxantrone that is complexed to the alginate through the ions will be released after the dissolution of the gel[30]. Amphiphilic alginate gels were used for modulating the hydrophobic drug release. Poor water soluble drug, theophylline was delivered in controlled manner by using the alginate grafted with poly (ε-caprolactone)[31] or the alginate incorporated with carbon nanotubes[32]. Alginate multi-particulate systems show faster drug release in simulated enteric environment compared to that of gastric environment[33]. The passive targeting of the drug, albendazole to the gastro intestinal tract was achieved using magnetic alginate beads[34]. The alginate hydrogels that incorporated with amoxicillin loaded nanoparticles were used for the effective treatment of *Helicobacter pylori* infections[35].

Alginate is a very good system for the delivery of protein based drugs since proteins can be incorporated into the alginates under non-denaturing conditions and they are prevented from degradation till their release from the gels. Normally, proteins will be quickly released from alginate since alginate gels are hydrophilic and porous. The protein release from the alginate gels can be modulated by changing the rate of degradation of gel. For example, the alginate that oxidized partially was used to make VEGF (vascular endothelial growth factor) release dependent on the degradation of alginate[36,37]. Proteins with high pI like lysozyme and chymotrypsin will physically cross-link with sodium alginate and get released in a sustained manner[38]. Alginate derivative obtained after reacting with amino group-terminated poly((2-dimethylamino) ethyl methacrylate) was used for the delivery of proteins orally[39]. Alginate gels could protect the acid labile insulin from denaturation at acidic pH and the protein could be released at neutral pH[40]. However, many proteins show lower encapsulation efficiency and faster release with alginate gels. These problems were addressed by modifying cross-linking or encapsulation strategies and improving the interactions between protein and hydrogel. Few examples are listed below. Alginate was blended with anionic polymers and then coated with chitosan for protecting insulin at gastric pH and releasing in a sustained manner at intestinal pH[41]. Alginate microspheres were coated layer by layer with silk fibroin in order to provide mechanical stability and diffusion barrier to the encapsulated proteins[42]. Alginate hydrogels loaded with PLGA microspheres were used for the controlled release of heat shock protein 27 that fused with transcriptional activator[43,44].

Alginate gels have been also explored for tissue engineering and regeneration. In these applications, alginates will be used for the delivery of cells or proteins that play role in the regeneration of tissues or organs. The factors that play an important role in the formation of blood vessels are VEGF and PDGF (plate derived growth factor). Angiogenesis and new capillary formation will be initiated by VEGF and maturation of capillaries in to functional blood vessels will be promoted by PDGF. Two strategies were reported in order to obtain sequential release of VEGF and PDGF from the alginate gels. In the first strategy, alginate gel with free VEGF was encapsulated with the microspheres that obtained after pre-encapsulation of PDGF into poly(lactide-co-glycolide)[45]. In another approach, the higher binding of PDGF to heparin was explored to retard its release compare to VEGF. In this strategy, free forms of both the factors were simply encapsulated to the alginate gel[46]. In either case, PDGF released slower than VEGF. The attractive approach for generating new blood vessel is cell transplantation, when the host cells won't respond to delivered growth factors. However, this approach was not promising when endothelial cells or their progenitor cells alone were used. But, when endothelial cells were transplanted with simultaneous delivery of VEGF from the alginate gel, new blood vessels were formed[47] and dual delivery of monocyte chemotactic protein-1 and VEGF through the alginate vehicles along with endothelial cells increased the number of matured vessels in the mice[48].

Alginate gels are helpful in the regeneration of bones by delivering osteoinductive factors, regenerating cells or both. These gels have the following advantages during bone or cartilage regeneration. They can fill defects of irregular shapes, can be easily modified with adhesive ligands like RGD, less invasive and they can release the induction factors in a controlled manner. Alginate gels also have the disadvantages like, weak mechanical properties and non-degradable inherent nature under physiological conditions. Bone tissue was regenerated by using the alginate gels that can deliver the DNA that encode for the bone morphogenetic proteins (BMPs)[49,50]. Dual delivery of BMP-2 and VEGF using alginate gels, improved the repair of bone defects which are at critical stage[51]. When primary rat osteoblasts were transplanted in to mice using RGD-alginate gels, bone formation was enhanced *in vivo*[52]. Degradable-alginate derived gels were also used for the regeneration of bone tissues[53]. Mesenchymal stem cells and BMP-2 entrapped alginate-chitosan gels lead to trabecular bone regeneration in the mice[54]. Chondrocytes suspended in the alginate solution (mixed with calcium sulfate) were used for the regeneration of cartilage[55,56]. Human mesenchymal stem cells that encapsulated in the alginate beads were grown in the medium that contains TGF-β1, ascorbate 2-phosphate and dexamethasone but not serum. These cells were shown to form cartilage in osteochondral defects[57].

Strategies for the regeneration of skeletal muscle comprise delivery of growth factors, cell transplantation or both[58,59]. The sustained and localized delivery of insulin like growth factor-1 and VEGF from alginate gels led to the significant regeneration of muscles[60]. Migration of primary myoblasts from RGD-alginate gels into damaged muscle *in vivo* was significantly increased with the simultaneous delivery of hepatocyte growth factor and fibroblast growth factor-2 from the gels[61]. This led to the enhanced regeneration of muscle fibers at the wounds[62]. Anisotropic gels based on the alginate didn't evoke major inflammatory response and promoted the regeneration of axons, when they were introduced into the acute lesions of spinal cord in adult rats[63]. Neural stem cells derived from the mouse were grown in the beads of calcium alginate and these cells could able to differentiate into neurons and glial cells[64]. Hence, alginate gels can be explored for the cell based neural therapy. Primary hepatocytes of the rat were viable in the alginate gels and these cells could synthesize fibronectin that in turn promoted the functional expression of spheroids[65]. The alginate gels that can deliver VEGF could improve the hepatocyte engraftment in the liver lobe of the rats[66]. In an effort to treat type 1 diabetes, alginate gels were also explored for transplanting the pancreatic islet. In this method, the immunosuppressive drugs that required to prevent graft rejection were avoided since alginate gel will protect the graft from the immune system of the host. This approach was successful in animal models[67,68,69]. Islet encapsulated alginate beads will be coated with poly (amino acids) like poly-L-lysine in order to reduce the size of outer pores and to retain the liquid core[70].

Alginate dressings will provide the physiologically wet environment and facilitate the healing of wounds[71]. Partially oxidized alginate gels accelerated the wound healing by releasing the dibutyryl cyclic adenosine monophosphate (regulator of human keratinocyte proliferation) in a sustained manner[72]. Alginate gels that encapsulated with stromal cell-derived factor-1 were effective in accelerating wound healing[73]. Incorporation of silver and zinc ions into the alginate dressings enhanced the anti-microbial effects at the site of wound[74,75]. Alginate-capped amphotericin-B lipid nanoconstructs act more effectively against leishmania parasite compared to the amphotericin-B lipid nanoconstructs alone[76]. Alginate-poly-L-lysine microparticles are proposed to be the potential carriers for antisense oligonucleotides[77]. Alginate-sterculia gum beads were used for retaining the anti-ulcer drug, pantoprazole for longer time in the stomach[78].

Conclusion

Alginate and its derivatives were widely explored as delivery systems for small drugs, nucleic acids and therapeutic proteins. The favourable properties of alginate for its biomedical applications include mild conditions for gelation, biocompatibility and safety in clinical use. However, the mechanical stiffness of the alginate gels is limited and this poses a challenge while matching their physical properties to the requirements of a particular application. This challenge can be overcome by adopting various cross-linking strategies that described earlier, blending with other molecules and using alginates of different molecular weight. Un-reacted reagents should be thoroughly removed from the alginate gels before using them in the delivery applications in order to avoid the non-specific reactions inside the host. As described earlier, alginate gels were used for the delivery of single or multiple drugs in a sustained or sequential manner. More research towards the precise control over the delivery of drugs from the alginate gels will improve the effectiveness and safety of the drugs. Studies that focus on the generation or synthesis of novel alginate polymers that have specific and better properties for the particular application will have significant impact on the use of alginate in the biomedical applications. Genetic engineering of the organisms that produce alginate might lead to the production of alginates with novel properties.

Acknowledgement

Authors are thankful to Dr. Chenraj Jain, Jain University Trust for his encouragement, support and facilities.

References

1. Tanaka, H.; Matsumara, M.; Veliky, I.A.; Diffusion characteristics of substrates in Ca-alginate beads, Biotechnology and Bioengineering, 1984, 26, 53-58.
2. Smith, T.J.; Calcium alginate hydrogel as a matrix for enteric delivery of nucleic acids, Biopharma, 1994, 4, 54-55.
3. Sutherland, I.W.; Alginates, in: D. Byrom (Ed.), Biomaterials; Novel Materials from Biological Sources, Stockton, New York, 1991, 309-331.
4. Black, W.A.P.; Woodward, F.N.; Alginates from common British brown marine algae. In Natural plant hydrocolloids. Adv.Chem.Ser.Am.Chem. Soc, 1954, 11, 83-91.
5. Zimmermann, U.; Klöck, G.; Federlin, K.; Hannig, K.; Kowalski, M.; Bretzel, R.G. et al., Production of mitogen-contamination free alginates with variable ratios of mannuronic acid to guluronic acid by free flow electrophoresis, Electrophoresis, 1992,13, 269-74.
6. Klöck, G.; Frank, H.; Houben, R.; Zekorn, T.; Horcher, A.; Siebers, U et al., Production of purified alginates suitable for use in immunoisolated transplantation, Appl Microbiol Biotechnol, 1994, 40, 638-43.
7. Haug, A.; Larsen, B.; Smidsrod, O.; A study of the constitution of alginic acid by partial acid hydrolysis, Acta Chem.Scand, 1966, 20, 183-90.
8. Haug, A.; Larsen, B.; Smidsrod, O.; Uronic acid sequence in alginate from different source, Carbohydr.Res, 1974, 32, 217-25.
9. Grasdalen, H.; Larsen, B.; Smidsrod, O.; ^{13}C-NMR studies of monomieric composition and sequence in alginate, Carbohydr Res, 1981, 89, 179-91.
10. Siddhesh, N Pawar.; Kevin, J. Edgar.; Alginate derivatization: A review of chemistry, properties and applications, Biomaterials, 2012, 33, 3279-3305.
11. LeRoux, M.A.; Guilak, F.; Setton, L.A.; Compressive and shear properties of alginate gel: effects of sodium ions and alginate concentration, J Biomed Mater Res, 1999, 47, 46-53.
12. Kong, H.J.; Smith, M.K.; Mooney, D.J.; Designing alginate hydrogels to maintain viability of immobilized cells, Biomaterials, 2003, 24, 4023-9.
13. Lee, K.Y.; Mooney, D.J.; Alginate: properties and biomedical applications, Prog Polym Sci, 2012, 37, 106-26.
14. Timell, T.E.; The acid hydrolysis of glycosides: I. General conditions and the effect of the nature of the aglycone, Can J Chem, 1964, 42, 1456.
15. Haug, A.; Larsen, B.; Phenolic compounds in brown algae: I. The presence of reducing compounds in Ascophyllum nodosum (L.), Le Jol. Acta Chem Scand, 1958, 12, 650.

16. Smidsrød, O.; Haug, A.; Larsen, B.; The influence of reducing substances on the rate of degradation of alginates, Acta Chem Scand, 1963, 17, 1473-4.
17. Tsujino, I.; Saito, T.; A new unsaturated uronide isolated from alginase hydrolysate, Nature, 1961,192, 970-1.
18. Preiss, J.; Ashwell, G.; Alginic acid metabolism in bacteria, J Biol Chem, 1962, 237, 309-16.
19. Smidsrød, O; Haug, A; Larsen, B; Degradation of alginate in the presence of reducing compounds, Acta Chem Scand, 1963, 17, 2628-37.
20. Leo, W.J.; Mcloughlin, A.J.; Malone, D.M.; Effects of sterilization treatments on some properties of alginate solutions and gels, Biotechnol Prog, 1990, 6, 51-3.
21. Otterlei, M.; Ostgaard, K.; Skjakbraek, G.; Smidsrod, O.; Soonshiong, P.; Espevik, T. Induction of cytokine production from human monocytes stimulated with alginate. J Immunother 1991, 10, 286-91.
22. Orive, G.; Ponce, S.; Hernandez, R.M.; Gascon, A.R.; Igartua, M.; Pedraz, J.L.; Biocompatibility of microcapsules for cell immobilization elaborated with different type of alginates. Biomaterials 2002, 23, 3825-31.
23. Lee, J.; Lee, K.Y.; Local and sustained vascular endothelial growth factor delivery for angiogenesis using an injectable system. Pharm Res 2009, 26, 1739-44.
24. Chickering, D.E; Mathiowitz, E; Bioadhesive Microspheres: A Novel Electro-Based Method to Study Adhesive Interactions Between Individual Microspheres and Intestinal Mucosa, J. Control Release 34, 1995, 251-261.
25. Grant, G.T; Morris, E.R; Rees, D.A; Smith, P.J.C; Thom, D; Biological interactions between polysaccharides and divalent cations – egg-box model, FEBS Lett, 1973, 32, 195-8.
26. Lee, K.Y; Kong, H.J; Larson, R.G; Mooney, D.J; Hydrogel formation via cell cross-linking, Adv Mater, 2003, 15, 1828-32.
27. Drury, J.L.; Boontheekul, T.; Mooney, D.J.; Cellular cross-linking of peptide modified hydrogels, J Biomech Eng: Trans ASME, 2005, 127, 220-8.
28. Tonnesen, H. H.; and Karlsen, J.; Alginate in drug delivery systems, drug development and industrial pharmacy, 2002, 28(6), 621-630.
29. Boontheekul, T.; Kong, H.J.; Mooney, D.J.; Controlling alginate gel degradation utilizing partial oxidation and bimodal molecular weight distribution, Biomaterials 2005, 26, 2455-65.
30. Bouhadir, K.H.; Alsberg, E.; Mooney, D.J.; Hydrogels for combination delivery of antineoplastic agents, Biomaterials, 2001, 22, 2625-33.

31. Colinet, I.; Dulong, V.; Mocanu, G.; Picton, L.; Le, Cerf D.; New amphiphilic and pH-sensitive hydrogel for controlled release of a model poorly water-soluble drug, Eur J Pharm Biopharm, 2009, 73, 345-50.
32. Zhang, X.L.; Hui, Z.Y.; Wan, D.X.; Huang, H.T.; Huang, J.; Yuan, H.; Yu, J.H.; Alginate microsphere filled with carbon nanotube as drug carrier, Int J Biol Macromol, 2010, 47, 389-95.
33. Lucinda-Silva, R.M.; Salgado, H.R.N.; Evangelista, R.C.; Alginate-chitosan systems: *in vitro* controlled release of triamcinolone and *in vivo* gastrointestinal transit, Carbohydr Polym 2010; 81: 260-8.
34. Wang, F.Q.; Li, P.; Zhang, J.P.; Wang, A.Q.; Wei, Q.; A novel pH-sensitive magnetic alginate74 chitosan beads for albendazole delivery, Drug Dev Ind Pharm 2010, 36, 867-77.
35. Chang, C.H.; Lin, Y.H.; Yeh, C.L.; Chen, Y.C.; Chiou, S.F.; Hsu, Y.M.; Chen, Y.S.; Wang, C.C.; Nanoparticles incorporated in pH-sensitive hydrogels as amoxicillin delivery for eradication of Helicobacter pylori, Biomacromolecules, 2010, 11, 133-42.
36. Lee, K.Y.; Peters, M.C.; Mooney, D.J.; Comparison of vascular endothelial growth factor and basic fibroblast growth factor on angiogenesis in SCID mice, J Control Release, 2003, 87, 49-56.
37. Silva, E.A.; Mooney, D.J.; Effects of VEGF temporal and spatial presentation on angiogenesis, Biomaterials, 2010, 31, 1235-41.
38. Wells, L.A.; Sheardown, H.; Extended release of high pI proteins from alginate microspheres via a novel encapsulation technique, Eur J Pharm Biopharm, 2007, 65, 329-35.
39. Gao, C.M.; Liu, M.Z.; Chen, S.L.; Jin, S.P.; Chen, J; Preparation of oxidized sodium alginate- graft poly((2-dimethylamino) ethyl methacrylate) gel beads and *in vitro* controlled release behavior of BSA, Int J Pharm, 2009, 371, 16-24.
40. Chan, A.W.; Neufeld, R.J.; Tuneable semi-synthetic network alginate for absorptive encapsulation and controlled release of protein therapeutics. Biomaterials 2010; 31, 9040-7.
41. Silva, C.M.; Ribeiro, A.J.; Ferreira, D.; Veiga, F.; Insulin encapsulation in reinforced alginate microspheres prepared by internal gelation, Eur J Pharm Sci, 2006, 29, 148-59.
42. Wang, X.; Wenk, E.; Hu, X.; Castro, G.R.; Meinel, L.; Wang, X.; Li, C.; Merkle, H.; Kaplan, D.L.; Silk coatings on PLGA and alginate microspheres for protein delivery, Biomaterials, 2007, 28, 4161-9.
43. Lee, J.; Lee, K.Y.; Injectable microsphere/hydrogel combination systems for localized protein delivery, Macromol Biosci, 2009, 9, 671-6.

44. Lee, J.; Tan, C.Y.; Lee, S.K.; Kim, Y.H.; Lee, K.Y.; Controlled delivery of heat shock protein using an injectable microsphere/hydrogel combination system for the treatment of myocardial infarction, J Control Release, 2009, 137, 196-202.
45. Sun, Q.H.; Silva, E.A.; Wang, A.X.; Fritton, J.C.; Mooney, DJ.; Schaffler, M.B.; Grossman, P.M.; Rajagopalan, S.; Sustained release of multiple growth factors from injectable polymeric system as a novel therapeutic approach towards angiogenesis, Pharm Res, 2010, 27, 264-71.
46. Hao, X.J.; Silva, E.A.; Mansson-Broberg, A.; Grinnemo, K.H.; Siddiqui, A.J.; Dellgren, G.; Wardell, E.; Brodin, L.A.; Mooney, D.J.; Sylven, C.; Angiogenic effects of sequential release of VEGFA(165) and PDGF-BB with alginate hydrogels after myocardial infarction, Cardiovasc Res, 2007, 75, 178-85.
47. Peters, M.C.; Polverini, P.J.; Mooney, D.J.; Engineering vascular networks in porous polymer matrices, J Biomed Mater Res, 2002, 60, 668-78.
48. Jay, S.M.; Shepherd, B.R.; Andrejecsk, J.W.; Kyriakides, T.R.; Pober, J.S.; Saltzman, W.M.; Dual delivery of VEGF and MCP-1 to support endothelial cell transplantation for therapeutic vascularization, Biomaterials 2010; 31: 3054-62.
49. Lopiz-Morales, Y.; Abarrategi, A.; Ramos, V.; Moreno-Vicente, C.; Lopez-Duran, L.; Lopez- Lacomba, J.L.; Marco, F.; *In vivo* comparison of the effects of rhBMP-2 and rhBMP-4 in osteochondral tissue regeneration, Eur Cells Mater, 2010, 20, 367-78.
50. Krebs, M.D.; Salter, E.; Chen, E.; Sutter, K.A.; Alsberg, E.; Calcium alginate phosphate-DNA nanoparticle gene delivery from hydrogels induces *in vivo* osteogenesis, J Biomed Mater Res Part A, 2010, 92, 1131-8.
51. Kanczler, J.M.; Ginty, P.J.; White, L.; Clarke, N.M.P.; Howdle, S.M.; Shakesheff, K.M.; Oreffo, R.O.C.; The effect of the delivery of vascular endothelial growth factor and bone morphogenic protein-2 to osteoprogenitor cell populations on bone formation. Biomaterials, 2010, 31, 1242-50.
52. Alsberg, E.; Anderson, K.W.; Albeiruti, A.; Franceschi, R.T.; Mooney, D.J.; Cell-interactive alginate hydrogels for bone tissue engineering, J Dental Res, 2001, 80, 2025-9.
53. Lee, K.Y.; Alsberg, E.; Mooney, D.J.; Degradable and injectable poly(aldehyde guluronate) hydrogels for bone tissue engineering, J Biomed Mater Res, 2001, 56, 228-33.
54. Park, D.J.; Choi, B.H.; Zhu, S.J.; Huh, J.Y.; Kim, B.Y.; Lee, S.H.; Injectable bone using chitosan alginate gel/mesenchymal stem cells/BMP-2 composites, J Cranio Maxill Surg, 2005, 33, 50-4.

55. Chang, S.C.N.; Rowley, J.A.; Tobias, G.; Genes, N.G.; Roy, A.K.; Mooney, D.J.; Vacanti, C.A.; Bonassar, L.J.; Injection molding of chondrocyte/alginate constructs in the shape of facial implants, J Biomed Mater Res, 2001, 55, 503-11.

56. Chang, S.C.N.; Tobias, G.; Roy, A.K.; Vacanti, C.A.; Bonassar, L.J.; Tissue engineering of autologous cartilage for craniofacial reconstruction by injection molding, Plast Reconstr Surg, 2003, 112, 793-9.

57. Ma, H.L.; Hung, S.C.; Lin, S.Y.; Chen, Y.L.; Lo, W.H.; Chondrogenesis of human mesenchymal stem cells encapsulated in alginate beads, J Biomed Mater Res Part A 2003, 64, 273-81.

58. Saxena, A.K.; Marler, J.; Benvenuto, M.; Willital, G.H.; Vacanti. J.P.; Skeletal muscle tissue engineering using isolated myoblasts on synthetic biodegradable polymers: preliminary studies, Tissue Eng, 1999, 5, 525-32.

59. Levenberg, S.; Rouwkema, J.; Macdonald, M.; Garfein, E.S.; Kohane, D.S.; Darland, D.C.; Marini, R.; van Blitterswijk, C.A.; Mulligan, R.C.; D'Amore, P.A.; Langer, R.; Engineering vascularized skeletal muscle tissue, Nat Biotechnol, 2005, 23, 879-84.

60. Borselli, C.; Storrie, H; Benesch-Lee, F.; Shvartsman, D.; Cezar, C.; Lichtman, J.W.; Vandenburgh, H.H.; Mooney, D.J.; Functional muscle regeneration with combined delivery of angiogenesis and myogenesis factors, Proc Natl Acad Sci USA, 2010, 107, 3287-92.

61. Hill, E.; Boontheekul, T.; Mooney, D.J.; Designing scaffolds to enhance transplanted myoblast survival and migration, Tissue Eng, 2006, 12, 1295-304.

62. Hill, E.; Boontheekul, T.; Mooney, D.J.; Regulating activation of transplanted cells controls tissue regeneration. Proc Natl Acad Sci USA, 2006, 103, 2494-9.

63. Prang, P.; Muller, R.; Eljaouhari, A.; Heckmann, K.; Kunz, W.; Weber, T.; Faber, C.; Vroemen, M.; Bogdahn, U.; Weidner, N.; The promotion of oriented axonal regrowth in the injured spinal cord by alginate-based anisotropic capillary hydrogels, Biomaterials, 2006, 27, 3560-9.

64. Li, X.Q.; Liu, T.Q.; Song, K.D.; Yao, L.S.; Ge, D.; Bao, C.Y.; Ma, X.H.; Cui, Z.F.; Culture of neural stem cells in calcium alginate beads, Biotechnol Prog, 2006, 22, 1683-9.

65. Glicklis, R.; Shapiro, L.; Agbaria, R.; Merchuk, J.C.; Cohen, S.; Hepatocyte behavior within three dimensional porous alginate scaffolds, Biotechnol Bioeng, 2000, 67, 344-53.

66. Kedem, A.; Perets, A.; Gamlieli-Bonshtein, I.; Dvir-Ginzberg, M.; Mizrahi, S.; Cohen, S.; Vascular endothelial growth factor-releasing scaffolds enhance vascularization and engraftment of hepatocytes transplanted on liver lobes, Tissue Eng, 2005, 11, 715-22.

67. Lim, F.; Sun, A.M.; Microencapsulated islets as bioartificial endocrine pancreas. Science 1980, 210, 908-10.
68. Lacy, P.E.; Hegre, O.D.; Gerasimidivazeou, A.; Gentile, F.T.; Dionne, K.E.; Maintenance of normoglycemia in diabetic mice by subcutaneous xenografts of encapsulated islets, Science, 1991, 254, 1782-4.
69. Calafiore, R.; Alginate microcapsules for pancreatic islet cell graft immunoprotection: struggle and progress towards the final cure for type 1 diabetes mellitus, Exp Opin Biol Ther, 2003, 3, 201-5.
70. Sakai, S.; Ono, T.; Ijima, H.; Kawakami, K.; *In vitro* and *in vivo* evaluation of alginate/sol- gel synthesized aminopropyl silicate/ alginate membrane for bioartificial pancreas. Biomaterials 2002, 23, 4177-83.
71. Queen, D.; Orsted, H.; Sanada, H.; Sussman, G.; A dressing history, Int Wound J 2004, 1, 59-77.
72. Balakrishnan, B., Mohanty, M.; Fernandez, A.C.; Mohanan, P.V; Jayakrishnan, A.; Evaluation of the effect of incorporation of dibutyryl cyclic adenosine monophosphate in an in situforming hydrogel wound dressing based on oxidized alginate and gelatin, Biomaterials, 2006, 27, 1355-61.
73. Rabbany, S.Y.; Pastore, J.; Yamamoto, M.; Miller, T.; Rafii, S.; Aras, R.; Penn, M.; Continuous delivery of stromal cell-derived factor-1 from alginate scaffolds accelerates wound healing, Cell Transplant, 2010, 19, 399-408.
74. Wiegand, C.; Heinze, T.; Hipler, U.C.; Comparative *in vitro* study on cytotoxicity, antimicrobial activity, and binding capacity for pathophysiological factors in chronic wounds of alginate and silver-containing alginate, Wound Repair Regener, 2009, 17, 511-21.
75. Agren, M.S. Zinc in wound repair, Arch Dermatol, 1999, 135, 1273-4.
76. Singodia, D.; Khare, P.; Dube, A.; Talegaonkar, S.; Khar, R.K.; Mishra, P.R.; Development and performance evaluation of alginate-capped amphotericin B lipid nanoconstructs against visceral leishmaniasis, J Biomed Nanotecnol, 2011, 7(1), 123-4.
77. González, F. M.; Tillman, L.; Hardee. G.; Bodmeier. R.; Characterization of alginate/poly-L-lysine particles as antisense oligonucleotide carriers, Int J Pharm, 2002, 4, 239(1-2), 47-59.
78. Singh, B.; Sharma, V.; Chauhan, D.; Gastroretentive floating sterculia–alginate beads for use in antiulcer drug delivery, Chemical Engineering Research, 2010, 88(8), 997-1012.

7 Polysaccharides used for Microencapsulation Processes

Leilane Costa de Conto[1] and Gustavo Henrique Santos Flores Ponce[2]

[1]*Professor, Federal Institute of Santa Catarina, Urupema, SC, Brazil* 88625-000.

[2]*Associate Professor, State University of Santa Catarina, Pinhalzinho, SC, Brazil* 89870-000.

7.1 Introduction

Microencapsulation is a technology that transforms solids, liquids or gases in solids, becoming a microscopic particle. Many microencapsulation techniques and different wall materials have been widely used for a variety of industrial, agricultural, medical, pharmacological and biotechnological applications. This chapter will discuss some researches and applications of polysaccharides that are being used to microencapsulate drugs.

7.2 Microencapsulation of Drugs

Microencapsulation is defined as the technology of packaging solids, liquids, or gaseous materials into sealed capsules with size of 1 to 1000 μm that can release their contents at controlled rates under specific conditions[1]. This concept was based on the idealized model of the cell in which the core is surrounded by a semipermeable membrane that protects it from external environment and, at the same time, controls the movemnet of substances in and out of cells[2].

The general microencapsulation purposes are: to facilitate its handling, transport and its addition in formulations; separate reactive materials; reduce toxicity of the active; sustain or prolong drug release; reduce volatility or flammability of liquids; mask taste and odor of certain components; increase the shelf life and protect against light, moisture and heat; prevent the incompatibility among the actives; avoid vaporization of many volatile compounds[3-5].

Microencapsulation technology finds its application in several industries, including pharmaceuticals for modified and controlled release of drugs. The controlled-release phenomenon has been documented for a vast array of drugs ranging from vasodilators, antihypertensives, bronchodilators to antibiotics. Table 7.1 shows the advantages and disadvantages of drugs controlled release.

TABLE 7.1

Drugs controlled release advantages and disadvantages

Advantages
Easy handling of the drug to the target organs and in the right amount at the right time;
Encapsulation capability of one or more drugs;
Many points of interactions with the bacteria's wall cell, as well as interactions with viruses;
Less toxicity (although the toxicity can be accentuated depending on the polymer wall and the drug encapsulated);
Changes or modifications of the drug activity by the change in solubility and diffusion rate;
Reduction of side effects such as irritation, nausea and vomiting, which are common in large doses of drugs;
Continuity of treatment during the night;
Protection of drugs which are rapidly destroyed by the body (this is particularly important in the macromolecules' release such as proteins which are now produced by genetic engineering);
Increasing the comfort of the patient;
Prolonged activity;
Drug delivery to a specific region of the body (lowering the systemic drug level);
Better patient adherence to treatment.
Disadvantages
Process costs and preparation of controlled release systems may be greater than the costs of the formulations patterns;
Risk of toxicity if the capsule is broken, which is accompanied by the release of large quantities of active;
Use of polymer additives such as antioxidants and stabilizers;
Risk related with the drug accumulation if the release rate is slow;
Low efficiency and lack of reproducibility, as a consequence of the individual desorption and metabolism;
Effects of the polymeric matrix in the environment;
Low adaption of the weaker drugs that require much larger quantities.

Source: Adapted 6.

Many techniques for encapsulation have been developed and their economic importance in industrial applications has resulted in considerable research, in special for pharmaceutical area. According to Gibbs et al.[7], these techniques may be divided into chemical and physical methods (simple or complex coacervation, organic phase separation, liposomal involvement), physical methods (spray drying, spray coating, spray chilling, fluid bed, extrusion, centrifugation with multiple orifices, co-crystallization, lyophilization) and chemical methods (interfacial polymerization, molecular inclusion).

The choice of method to encapsulate depends on the core material properties, especially solubility, the type of particle desired for protection and the controlled release, finality (morphology, stability and release mechanism) and the circumstances involved on manufacturing the product[8].

Another critical point in the microencapsulation is the selection of wall materials due to the active ingredient, the system on which microcapsule will be applied and its release mechanisms[9]. These materials may be lipids (wax, palm fat), proteins (gelatin, milk proteins, soy proteins), mono, di and oligosaccharides (hydrolyzed starches, lactose) and polysaccharides like gum Arabic, agar, alginate, carrageenan, starch, modified starches, dextrin, chitosan, saccharose, carboxymethyl celluloses, acetylcellulose, nitrocellulose, pectin among others[3,10]. These polysaccharides have gained an important role for gel entrapment and encapsulation, they possess unique gelling properties and offer a variety of gel formation mechanisms, thermal and chemical stability and ability to give crosslinked networks[11].

Because of the wide availability of encapsulated drugs, many products or functional ingredients whose development was thought to be technically unfeasible are now possible. Such ingredients are products of a process in which the active is enveloped in a coating or capsule, thereby conferring many useful or eliminating undesirable properties from the original drug. A list of microencapsulated drugs and their purposes is presented in Table 7.2.

TABLE 7.2

Examples of microencapsulated drugs and their purposes.

Drug material	Purpose
Acetaminophen	Taste masking
Aspirin	Taste masking, sustained release, reduce gastric irritation and incompatibilities

TABLE 7.2 ***Contd...***

Drug material	Purpose
Eprazinone	Easy handling and storage.
Islet of Langerhans	Sustained normalization of diabetic condition
Isosorbide dinitrate	Sustained release
Menthol	Reduction of volatility
Paracetamol, Nitrofurantoin etc	Bitter taste masking
Progesterone	Sustained release
Potassium chloride	Reduced gastric irritation
Urease	Perm selectivity of enzyme, substrate and reaction product
Vitamin A palmitate	Stabilization to oxidation
5-fluorouracil	Reduced irritation
Retinol	Enhanced potency and reduced irritation
Salicylic acid	Target particular area
Sodium chloride	Reduced hygroscopic properties

Source: [12,13]

Potential applications of drug delivery system are replacement of therapeutic agents (not taken orally today like insulin)[14,15] gene therapy[16-19], and in use of vaccines for treating AIDS[20-22], cancer[23], tumors[24,25] and diabetes[26-28]. Protein such as insulin, growth hormone[29] and erythropoietin[30,31], used to treat anemia, are example of drugs that would benefit from this new form of oral delivery. The delivery of corrective gene sequences in the form of plasmid DNA[32] could provide convenient therapy for a number of genetic diseases such as cystic fibrosis[33,34] and hemophilia[35].

According to Dubey et al.[36], based on this novel drug delivery technique, *Lupin* has already launched in the market worlds first Cephalexin (Ceff-ER) and Cefadroxil (Odoxil OD) antibiotic tablets for treatment of bacterial infections. Aspirin controlled release version ZORprin CR tablets are used for relieving arthritis symptoms. Quinidine gluconate CR tablets are used for treating and preventing abnormal heart rhythms. Niaspan CR tablet is used for improving cholesterol levels and thus reducing the risk for a heart attack. Glucotrol (Glipizide SR) is an anti diabetic medicine used to control high blood pressure.

7.3 Wall Materials for Microencapsulation

According to Bansode et al.[5], the coating material should be capable of forming a film that is cohesive with the core material; be chemically compatible and nonreactive with the core material; and provide the desired coating properties as: strength, flexibility, impermeability, optical properties and stability.

Several core materials are insoluble in aqueous solutions, then the formation of emulsions is necessary and becomes a significant property to consider in selecting an appropriate covering material[37]. Generally, proteins are the main emulsifiers and the polysaccharides contribute to the stability of emulsions through their thickening and steric stabilizing characteristics. The interactions between proteins and the carbohydrates can also help stabilizing the emulsions. Several polysaccharides have been used together with proteins to enhance the emulsion stability and microencapsulation. The combinations such as whey protein isolate/gum Arabic[38], β-lactoglobulin/pectin[39], β-lactoglobulin/κ-carrageenan[40], whey protein/chitosan/gum Arabic[41] and milk protein products/ xanthan[42] are frequently used.

The gums are usually interesting for industrial processing microcapsules. The term gum is used to refer to a group of polysaccharides or their derivatives obtained from plants (exudates, extracts, seed and algae) or substances secreted by bacteria[43]. The description of some gums utilized to microencapsulate drugs will be show in this section.

7.3.1 Gum Arabic

Gum Arabic (GA) is also called gum acacia is a complex polysaccharide which has three fractions of different molecular sizes, and one part protein corresponding to approximately 2% of the molecule, which confers characteristics emulsifiers and stabilizers[44].

It has a highly branched structure (Figure 7.1), presenting a flattened ellipsoid shape. The main chain consists of units 1.3 - β-D-galactopyranose joined by glycosidic bonds which are attached side chains with different chemical structures formed by D-galactopyranose, L-rhamnose, L-arabinose and D-glucuronic acid, linked to the main chain by links β (1.6). The natural polysaccharide is in salt form, which is converted by acidification to the corresponding acid, arabic acid. That is negatively charged at low pH (< 2.2) due to the dissociation of the carboxyl groups[45]. This peculiar composition confers to GA a charge density that is six times higher than a linear polysaccharide having the same composition and a good cold solubility because of the presence of residual charged groups and peptidic fragments. GA is a widely used polysaccharide in microencapsulation procedures, film formation and emulsion stabilisation[46].

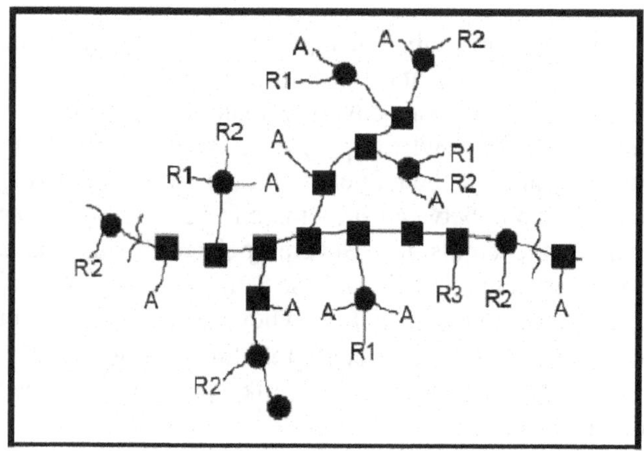

where: A = arabinosyl; based radicals R1 = rhamnose, acid gulucurônico R2 = R3 = arabinose. Source:[49]

FIGURE 7.1 Molecular structure of gum Arabic.

GA has good properties that allow its use in processes for encapsulation of lipids and volatile substances[8]. Its characteristics such as: high solubility, low viscosity at high concentrations, and emulsification characteristics and good capacity to encapsulation retention, filling the encapsulation of volatile compounds are derived from their low molecular weight, approximately 250,000 Da[47]. Although solutions containing up to 50% gum can be prepared, the solution viscosity starts to rise steeply at concentrations greater than 35%. Most other gums yield solutions with high viscosity at concentrations as low as 1%. These features make it very versatile for most microencapsulation processes, however, this wall material has a higher cost compared to other materials such as maltodextrin[48]. GA has advantages like its emulsifier property in a wide pH range, as well as creating a film around the droplets and good binding properties[8].

Several drugs have been microencapsulated by complex coacervation utilizing GA. Vitamin E, for instance, was coated by β-lactoglobulin/GA matrix, studied by Schmitt et al.[49] and Renard et al.[47]. Other vitamins have been used in the microcapsules, the matrix most studied for this was gelatin/GA by the complex coacervation. This matrix was used to microencapsulate vitamin A (β-carotene) as described by Gerritsen and Crum[50] and vitamin A palmitate as studied by Junyaprasert et al. [51]. For Vitamin C (ascorbic acid), Trindade and Grosso[52] studied the stability of ascorbic acid microencapsulated in granules of rice starch and in GA, it was concluded that, in terms of stability of microencapsulated ascorbic acid, the GA was the best coating in spray drying. Nixon and Agyilirah[53] worked with controlled release of Phenobarbitone. The traditional gelatin/GA system was investigated for the encapsulation of capsaicin, by Xing et al[54].

7.3.2 Alginate

Alginate is a group of unbranched polysaccharides composed of the epimers α-L-guluronic acid (G) and β-D-mannuronic acid (M) residues, arranged in homopolymeric blocks (MM, GG) and also in heteropolymeric blocks (MG), like exposed Figure 7.2. Alginates with a high M content are the best suited for thickening applications, whereas those with a high G content are best for gelation. Designer alginates are currently being developed, involving the 5-epimerization of β-(1→4)-linked M residues to α-(1→4)-linked G residues in algal alginates using bacterial epimerases[55].

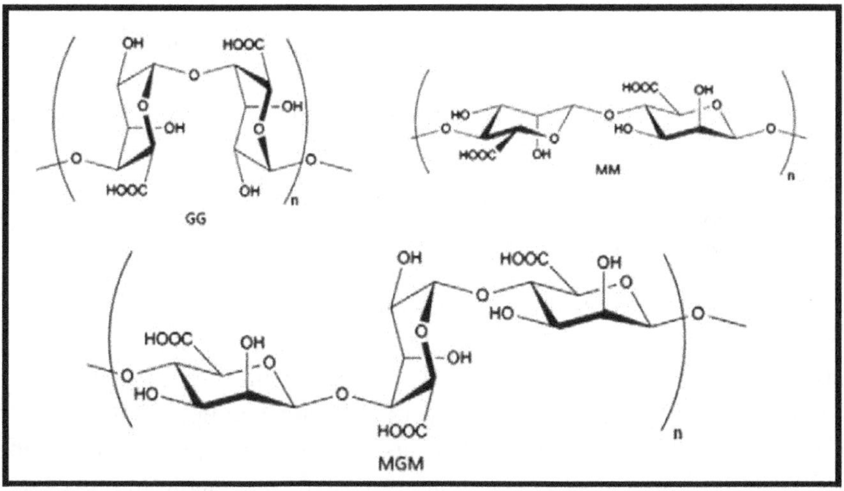

where: M = β-D-mannuronic acid unit; G = α-L-guluronic acid unit.
Source:[56]

FIGURE 7.2 Molecular structure of alginate.

Alginates form gels with divalent/multivalent cations (mainly Ca^{2+}) characterized by strength and elasticity. The M/G ratio is an important consideration in the process, when the ratio decreases, the requirement of Ca^{2+} for cross-linking increases, but monovalent cations and Mg^{2+} ions do not induce the gelation. The cations participate in the interchain binding between the G blocks to produce a three-dimensional network, often called an egg-box model. The gel network and homogeneity depends on the cation concentration. With an excess of Ca^{2+} a modified egg box having multiple alginate chains in the gelling zone may be formed with different physicochemical properties[57].

This ability has been used to produce microcapsules and cell immobilization by a process called ionotropic gelation. This involves the dropping of a concentrated alginate solution, most commonly through a needle, into calcium chloride solution, externally gelling the polymer into a microcapsule. The size of the microcapsules formed using external gelation is governed by the size of droplets formed during the extrusion process[58], particles from as little as tens of microns being produced by spray technology, up to millimeter size when needle extrusion is used. However, less commonly, the microcapsules may be formed by internal gelation, in which the alginate in solution contains calcium carbonate[59].

Numerous attributes make alginate an ideal drug-delivery vehicle. These features include a high water amount within the matrix, adhesive interactions with intestinal epithelium, a mild room temperature drug(s) encapsulation process free of organic solvents, a high gel porosity (allowing high diffusion rates of macromolecules), the ability to control this porosity with simple coating procedures using polycations and dissolution/biodegradation of the system under normal physiological conditions[60].

Hence, it is known that alginate has been used as a carrier for the controlled release of numerous molecules, for example, indomethacin[61], sodium diclofenac[62], nicardipine[55], dicoumarol[63], gentamicin[64], vitamin C, ketoconazole[65], amoxycillin[66] and antitubercular drugs[67]. George and Abraham[68] listed some studies about oral delivery of peptide or protein drugs utilizing alginate matrices by ionotropic gelation. Further, the encapsulation of peptides/DNA/cells in alginate is also likely to eliminate some of the staggering problems that face biotechnologists.

Another common use of alginate microcapsules is to reduce the viability losses of probiotics bacteria, like *Bifidobacterium* and *Lactobacillus*. Cook et al.[56] summarized some works on the alginate encapsulation of probiotics and, this work follows simplified in Table 7.3.

TABLE 7.3
Overview of some literature available on the alginate encapsulation of probiotics.

Encapsulation Material	Bacteria	Key findings
Alginate	Lactobacillus acidophilus	Capsule size affects viability in simulated gastric conditions
Alginate coated with palm oil and poly-L-lysine	8 different Lactobacilli and Bifidobacteria	Coating with palm oil and poly-L-lysine improved viability in simulated gastric conditions
Alginate and xanthan gum	Lactobacillus acidophilus	Encapsulation improved cell survival in simulated gastric juice at as low as pH 1.2
Alginate coated with either chitosan, alginate or poly-L-lysine-alginate	Lactobacillus acidophilus, Bifidobacterium bifidum, Lactobacillus casei	Chitosan improved survival of the bacteria best in simulated gastric juice.
Alginate	Lactobacillus casei	Bacterial survival during exposure to heat, bile salts and low pH, increases with alginate concentration.
Alginate, Alginate and pectin	Lactobacillus casei	Blending of pectin into microcapsules improved cell survival at low pH

7.3.3 Carrageenan

Carrageenan is a linear heteropolysaccharide with ester sulphate groups. It's main chain consists of alternating copolymers of 1,4- or 1,3-β-D-galactopyranose and 3,6-anhydro-D-galactopyranose[69]. Carrageenan can be found in three major forms that are designated κ (*kappa*), *i* (*iota*) and λ (*lambda*) that are exposed in Figure 7.3.

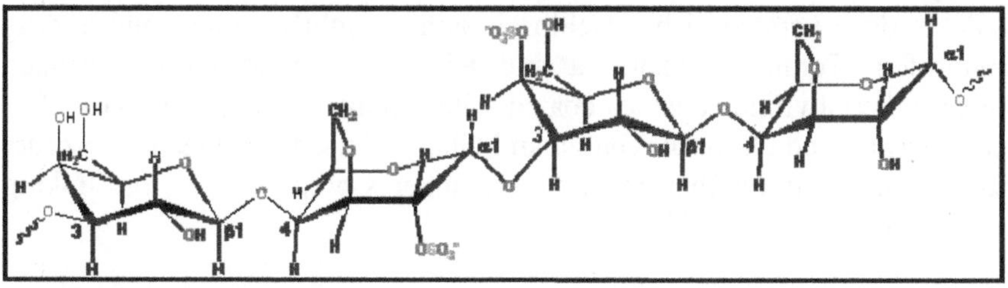

Source: [70]

FIGURE 7.3 The main carrageenan chain consists of alternating 3-linked-β-D-galactopyranose and 4-linked-D-galactopyranose units.

Carrageenans are highly flexible molecules which, even at higher concentrations, organize into double-helical structures. Gel formation in *kappa*- and *iota*-carrageenans involves helix formation on cooling from a hot solution together with gel-inducing and gel-strengthening; the helix–helix aggregation relies upon specific cation presence (Ca^{2+} and K^{+}), which can form screen electrostatic repulsive forces between the participating chains by packing within the aggregate structure[71].

Carrageenans are widely used in the pharmaceutical and other industries (cosmetic products, pesticides and food) as thickening and stabilizing agents. The use of carrageenans in the encapsulation field has been related in the literature. Previous studies have reported microcapsules produced by carrageenan and oligochitosan polymer[71], but the most reports are related to the use of carrageenan (in special *k*-carrageenan) for the encapsulation of microbial cells, due to the fact that its gelation occur with the change in temperature, between 40ºC to 45ºC, in order to preserve the sensitive materials[70].

Audet, Paquin, and Lacroix[72] reported the microencapsulation of some bacteria such as *Streptococcus thermophilus* and *Lactobacillus bulgaricus* using a combination of *k*-carrageenan and locust bean gum. Ouellette, Chevalier, and Lacroix[73] immobilized a pure culture of *Bifidobacterium infantis* in *k*-carrageenan/ locust bean gum beads. Even as Doleyres et al.[74] immobilized probiotic cells in carrageenan and locust bean gum gel beads by ionotropic gelation method to produce a mixed lactic culture containing a non-competitive strain of bifidobacteria.

Several studies have been performed to verify the delivery of microencapsulated drugs with carrageenan as wall material. Suzuki S, Lim JK.[75] microencapsulated gentamicin sulphate in carrageenan-locust bean gum mixture in a multiphase emulsification technique for sustained drug release. Sipahigil, and Dortunç[76] prepared and evaluated carrageenan beads as a controlled release system for a freely water soluble drug verapamil hydrochloride and a slightly water soluble drug ibuprofen. Tomida, Nakamura and Kiryu[77] worked with controlled release theophylline capsules coated with a polyelecrolyte complex of κ-carrageenan and chitosan. Garcıa and Ghaly[78] used carrageenan in order to control acetaminophen release from spheres prepared by cross linking technique.

7.3.4 Starch and Modified Starches

Starch is a polysaccharide that consists of a large number of glucose units joined together by glycosidic bonds. It comprises amylose and amylopectin as its macromolecules. Starch is found in corn, cassava, potatoes, wheat, rice and other foods, ranging in characteristics and appearance, depending on its source. In the native form, starches have limited use because it produce weak-bodied, cohesive, rubbery pastes when heated and undesirable gels when the pastes are cooled[79]. The properties of starches can be improved by various modifications methods.

Researchers have developed methods to modify starch, which requires mostly the usage of enzymes and chemicals. Modified starch is obtained by a partially degraded starches or addiction of some chemical groups. The purposes of this modification are to enhance its properties in specific applications, such as to improve the increase in water holding capacity, heat resistant behavior, reinforce its binding, minimized syneresis and improved thickening[79,80].

Enzymatic modification of starch consist in a hydrolysis of some part of starch, reducing into a low molecular weight of starch called maltodextrins, or dextrin using amylolytic enzymes[80]. These types of modified starches are widely used for food and pharmaceutical industries.

Chemical modification is the mainstream of the modified starch in the last century. Many developments on chemical modification of starches have been introduced in food, pharmaceutical and textile industries.

Starches and their derivatives have been applied to the micro-encapsulation of vitamins. Stability of ascorbic acid microencapsulated with starch or modified starches were studied by Trindade and Grosso[52], Uddin et al. [81] and Dib Taxi et al.[82].

Certain studies have been turning to the microencapsulated drug release in starch and its derivatives matrices, such as: propranolol HCl encapsulated by interfacial cross-linking of β-cyclodextrins with terephthaloylchloride[83]; protein and a proteinase inhibitor in order to utilizing starch and bovine serum albumin as wall material[84]; Ubiquinone-10 (CoQ10) microencapsulated using different blends of gum arabic, maltodextrin and modified starches as wall materials[85].

7.3.5 Chitosan

Chitosan is a cationic polymer, that the most commonly obtained form is the α-chitosan from crustacean chitin (crab and shrimp shell wastes). In preparing chitosan, ground shells are deproteinated and demineralized by sequential treatment with alkali and acid. After that, the extracted chitin is deacetylated to chitosan by alkaline hydrolysis at high temperature[68].

This polysaccharide is a linear co polymer consisting of β (1–4)-linked 2-amino-2-deoxy-D-glucose (D-glucosamine) and 2-acetamido-2-deoxy-D-glucose (N-acetyl-D-glucosamine) units (Figure 7.4). The structure of chitosan is very similar to that of cellulose (made up of β (1–4)-linked D-glucose units), in which there are hydroxyl groups at C2 positions of the glucose rings[86].

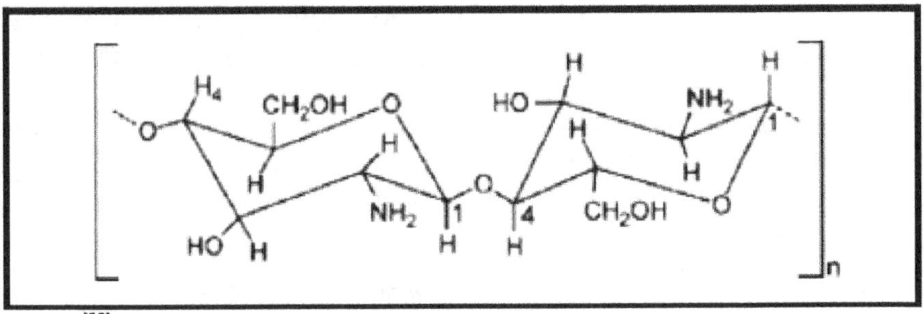

Source:[68]

FIGURE 7.4 Molecular structure of chitosan.

The properties, biodegradability and biological role of chitosan is frequently dependent on the relative proportions of N-acetyl-D-glucosamine and D-glucosamine residues, having an excellent film-forming ability. Chitosan films exhibit limited swelling in water. However, through the formation of blends and interpenetrated or semi-interpenetrated polymer networks with highly hydrophilic polymers such as poly(vinyl alcohol), polyether, or polyvinylpyrrolidone, membranes with varying degrees of hydrophilicity have been obtained and used as matrices for different applications[87].

The gelling properties of chitosan allow its use for a wide range of applications, the most attractive are coating of foods and pharmaceuticals. Others attractive are gel entrapment of biochemicals, plant embryos and whole cells, microorganisms, or algae. Such entrapment offers diverse uses including microencapsulation and controlled release of flavors, nutrients or drugs. Because chitosan has been shown effectiveness, concurrent cell permeabilization and immobilization using chitosan-containing complexes of coacervate capsules have been widely explored[86].

There are numerous scientific reports and patents on the preparation of chitosan microspheres and microcapsules, including some reviews on the subject[87]. The techniques employed to encapsulate with chitosan include: ionotropic gelation; spray drying; emulsion phase separation;

simple and complex coacervation. chitosan coatings; among others[88]. Combinations of the techniques above are also used in order to obtain microparticles with specific properties and performances.

Chitosan is very used as drug carrier and some therapeutics purposes with respective drugs microencapsulated studied are presented in Table 7.4.

TABLE 7.4

Overview of some literature available on the chitosan encapsulation of drugs.

Therapeutic Purpose	**Drug**
Cancer Therapy	Gadolinium
	Doxorubicin
	5- Fluoruracil
	Metotrexate
	Adramycin
Therapeutic Purpose	**Drug**
Vaccine	Ovoalbumin
	Plasmid DNA
	Tetanus toxoid Protein
Diabetes	Insulin
Epilepsy	Fenobartinona
Antihypertensive	Diclofenac sodium
	Prednisolone Sódium Phosphate
	Ibuprofen
Ophthalmic drugs	Cyclosporine A
	Indomethacin
Osteoporosis	Salmon calcitonin

Source: Adapted[88]

7.3.6 Celluloses and Celluloses Derivatives

Cellulose is the main constituent of plant cell walls. It consists of glucopyranosyl residues joined by β-(1→4) linkages. Native cellulose is insoluble in water owing to the high level of intramolecular hydrogen bonding in the cellulose polymer. As an edible film for coatings, the permeability of cellulose can be modified by combining it with other coating materials via etherification. The procedure consist in reacting cellulose with aqueous caustic and then with methyl chloride, propylene oxide or sodium monochloroacetate to yield methylcellulose, hydroxypropyl methylcellulose, hydroxypropylcellulose and sodium carboxymethycellulose[89].

Carboxymethylcellulose (CMC) or commonly called cellulose gum® is a cellulose derivate with carboxymethyl groups bound to some of the hydroxyl groups of the glucopyranose monomers, that make up the cellulose structure (Figure 7.5). It is an anionic water-soluble polymer often used as its sodium salt, the sodium carboxymethylcellulose[90].

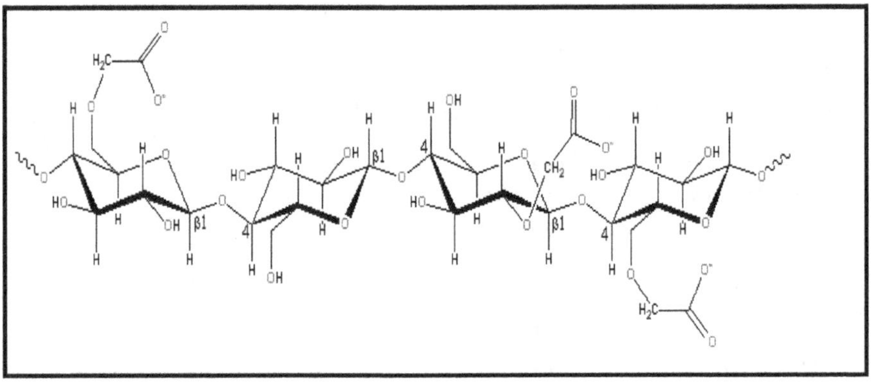

Source: (90)

FIGURE 7.5 Molecular structure of CMC.

CMC is used in a lot of industrials segments because of their specific features and properties. In solution CMC acts as: a thickener, suspending agent, stabilizer binder among others. CMC make resistant films, those are capable to resist to organic solvents, oils and greases for instance[89].

In the microencapsulation field, CMC and others cellulose derivates can be used. It was evidenced by the use as a coat material to drugs delivery like the examples found in the literature. Ahmad et al.[91] worked with microencapsulation of Diclofenac sodium employing different ratios of ethyl cellulose. Tiyaboonchai and Ritthidej[92] prepared a complex coacervation of chitosan (CS) and CMC (hardened with glutaraldehyde), to control the release of Indomethacin from microcapsule. The same active was encapsulated in ethyl cellulose (EC) and hydroxy propyl methyl cellulose phthalate (HPMCP) by o/w emulsification-solvent evaporation technique[93].

Other microencapsulation studies have been conducted utilizing cellulose derivates aiming to taste masking the bitter drugs: Cefuroximeaxetil with cellulose acetate phthalate and hydroxyl propyl methyl cellulose, Flucoxacillin with ethyl cellulose and with Azithromycin[94].

7.3.7 Others

Some polysaccharides are important for their ability to form films and as wall materials, however they have other more important applications just like pectin and agar that will be discussed briefly below.

Pectin is a water soluble anionic polysaccharide comprised predominantly of linear chains of (1→4)-linked α-D-galacturonic acid residues, which occur partially esterified with a small percentage of rhamnose units to yield branches consisting of neutral sugars, notably galactose and arabinose[95]. Pectin can form complexes with other polymers due to its charge balance, which presents positive at high pHs and negative at low pHs. Because of these characteristics electrostatic and gel formation, the pectin is presented as a good wall material used in microencapsulation processes. An example was the encapsulation of vitamin D in β-Lactoglobulin using a low methoxyl pectin (LMP) matrices[96].

Agar is a heterogeneous complex mixture of related polysaccharides having the same backbone chain structure. Its main component is β-D-galactopyranosyl linked (1→4) to a 3,6-anhydro-α-L-galactopyranosyl unit and partially esterified with sulfuric acid. Such structural regularity may be masked or modified in a number of ways by the substitution of hydroxyl groups out of sulphate hemiesters, as well as methyl ethers in various combinations and more rarely with a cyclic pyruvate ketal as 4,6-*O*-(1-carboxyethylidene) group[97].

Agar and agarose have been extensively used and are still widely used as stabilizing and gelling additives in food preparations. However, agar became best known as a thermoreversible ion-independent solidifier in bacteriology. They also have found application in pharmaceutical products as a thickener, suspending and gelling agent in formulations for controlled-release of drugs. In the field of biotechnology these have been used for immobilization of bacteria, yeasts and animal cells[11].

Some other polysaccharides are used for microencapsulation of drugs, such as gum locust, gum *ghatti*, gum *karaya*, gum *tragacanth*, although these are less studied and used due to the peculiar characteristics and shortage of supply.

Conclusion

This chapter presented the different polysaccharides (alone or interacting with others materials) that can be used as wall material to microencapsulation of drugs. The approach took into account the specific feature of each material, mostly related to microencapsulation purpose and how it can be obtained. A lot of examples were brought to this chapter in order to illustrate some applications of these materials to microencapsulate different drugs.

References

1. Augustin, M.A.; Sanguansri, L.; Margetts, C.; Young, B. Microencapsulation of food ingredients. Food Australia 2001, 56: 220-223.
2. Jizomoto, H.; Kanaoka, E.; Sugita, K.; Hirano, K. Gelatin-Acacia microcapsules for trapping micro oil droplets containing lipophilic dugs and ready disintegration in the gastrointestinal tract. Pharmaceutical Research 1993, 10 (8), 1115-1122.
3. Sanguansri, L.; Augustin, M. A. Microencapsulation and delivery of omega-3 fatty acids. Functional food ingredients and nutraceuticals. In Functional food ingredients and nutraceuticals; Shi, J., Ed.; CRC Press: Boca Raton, 2007; 297-327.
4. Favaro-Trindade, C.S.; Pinho, S.C.; Rocha, G.A. Revisão: Micro-encapsulação de ingredientes alimentícios. Brazilian Journal of Food Technology 2008, 11 (2), 103-112.
5. Bansode, S.S.; Banarjee, S.K.; Gaikwad, D.D.; Jadhav, S.L.; Thorat, R.M. Microencapsulation: A Review. International Journal of Pharmaceutical Sciences Review and Research 2010, 1 (2), 38-43.
6. Dumitriu, S.; Dumitriu, M. Polymeric Drug Carriers. In Polymeric Biomaterials; Dumitriu, S., Ed.; Marcel Dekker: New York, 1993; 447-452.
7. Gibbs, B.F.; Kermasha, S.; Alli, I.; Mulligan, C.N. Encapsulation in the food industry: a review. International Journal of Food Sciences and Nutrition 1999, 50, 213-224.
8. Lee, S.J.; Ying, D.Y. Encapsulation of fish oils. In Delivery and controlled release of bioactives food nutraceuticals; Garti, N., Ed.; CRC Press: Boca Raton, 2008; 370-403.
9. Davidov-Pardo, G.; Roccia, P.; Salgado, D.; León, A.E.; Pedroza-Islas, R. Utilization of different wall materials to microencapsulate fish oil evaluation of its behavior in bread products. American Journal of Technology 2008, 3 (6), 384-393.
10. Drusch, S.; Serfert, Y.; Scampicchio, M.; Schmidt-Hansberg, B.; Schwarz, K. Impact of physicochemical characteristics on the oxidative stability of fish oil microencapsulated by spray-drying. Journal of Agricultural and Food Chemistry 2007, 55, 11044-11051.
11. Murano, E. Use of natural polysaccharides in the microencapsulation techniques Journal of Applied Ichthyology 1998, 14, 245-249.

12. Jadupati, M.; Tanmay, D.; Souvik, G. Microencapsulation: an indispensable technology for drug delivery system. International Research Journal of Pharmacy 2012, 3 (4), 8-13.
13. Agnihotri, N.; Mishra, R.; Goda, C.; Arora, M. Microencapsulation – A Novel Approach in Drug Delivery: A Review. Indo Global Journal of Pharmaceutical Sciences 2012; 2 (1): 1-20.
14. Naha, P.C.; Kanchan, V.; Manna, P.K.; Panda, A.K. Improved bioavailability of orally delivered insulin using Eudragit-L 30D coated PLGA microparticles. Journal of Microencapsulation 2008, 25 (4), 248-256.
15. Kim, C.H.; Kwon, J.H.; Choi, S.H. Controlled Release preparation of Insulin and its method. US Patent 7,087,246 B2. 8 Aug 2006.
16. Jones, D.H.; Farrar, G.H.; Stephen, J.C. Microbiological Research Authority (GB). Method of making microencapsulated DNA for vaccination and Gene Therapy. US Patent 6,270,795. 7 Aug 2001.
17. Chang, P. L. Microencapsulation – An alternative approach to gene therapy. Transfusion and Apheresis Science 1996, 17 (1), 35-43.
18. Aihua, L.A.; Feng, S.; Tao, Z.; Pasquale, C.; Murray, P.; Patricia, C. Enhancement of myoblast microencapsulation for gene therapy. Journal of Biomedical Materials Research Part B: Applied Biomaterials 2006, 77 (2), 296-306.
19. Ross, C.J.D.; Ralph, M.; Chang, P.L. Somatic gene therapy for a neurodegenerative disease using microencapsulated recombinant cells. Experimental Neurology 2000, 166 (2), 276-286.
20. McMahon, J.; Schmid, S.; Weislow, O.; Stinson, S.; Camalier, R.; Gulakowski, R.; Shoemaker, R.; Kiser, R.; Harrison, S.; Mayo, J.; Boyd, M. Feasibility of cellular microencapsulation technology for evaluation of anti-human immunodeficiency virus *in vivo*. Journal of the National Cancer Institute 1990, 82 (22), 1761-1765.
21. Cleland, J.L.; Powell, M.F.; Lim, A.; Barron, L.; Berman, P.W.; Eastman, D.J.; Nunberg, J.H.; Wrin, T.; Vennari, J.C. Development of a single shot subunit vaccine for HIV-1. AIDS Research and Human Retroviruses 1994, 10 (2), 21-26.
22. Marx, P.A.; Compans, R.W.; Gettie, A.; Staas, J.K.; Gilley, R.M.; Mulligan, M.J.; Yamshchikov, G.V.; Chen, D.; Eldridge, J.H. Protection against vaginal SIV transmission with microencapsulated vaccine. Science 1993, 260 (5112), 1323-1327.
23. Drone, P.; Bourgeois, J.M.; Chang, P.L. Antiangeogenic cancer therapy with microencapsulated cells. Human Gene Therapy 2003, 14 (11), 1065-1077.

24. Hao, S.; Su, L.; Guo, X.; Moyana, T.; Xiang, J. A novel approach to tumor suppression using microencapsulated engineered J558/TNF-a cells. Experimental oncology 2005, 27 (1), 56-60.
25. Zhang, Y.; Wang, W.; Zhou, J.; Yu, W.; Zhang, X.; Guo, X.; Ma, X. Tumor anti-angiogenic gene therapy with microencapsulated recombinant CHO cells. Annals of Biomedical Engineering 2007, 35 (4), 605-614.
26. Maria-Engler, S.S.; Correa, M.L.C.; Oliveira, E.M.C.; Genzini, T.; Miranda, M.P.; Vilela, L.; Sogayar, M.C. Microencapsulation and tissue engineering as an alternative treatment of diabetes. Brazilian Journal of Medical and Biological Research 2001, 34 (6), 691-697.
27. Sambanis, A. Encapsulated islets in diabetes treatment. Diabetes Technology and Therapeutics 2003, 5 (4), 665-668.
28. Kizilel, S.; Wyman, J.L.; Mrksich, M.; Nagel, S.R.; Garfinkel, M.R. Brinks Hofer Gilson and Lione, US Patent 2007/0190036 A1. 16 Aug 2007.
29. Kim, H.K.; Park, T.W. Microencapsulation of human growth hormone within biodegradable polyester microspheres: Protein aggregation stability and incomplete release mechanism. Biotechnology and Bioengineering 1999, 65 (6), 659-667.
30. Morlock, M.; Koll, H.; Winter, G.; Kissel, T. Microencapsulation of rh-erythropoietin, using biodegradable poly(d,l-lactide-co-glycolide): protein stability and the effects of stabilizing excipients. European Journal of Pharmaceutics and Biopharmaceutics 1997, 43 (1), 29-36.
(31) Morlocka, M.; Kissela, T.; Lia, Y.X.; Kollb, H.; Winterb, G. Erythropoetin loaded microspheres prepared from biodegradable LPLG-PEO-LPLG triblock copolymers: protein stabilization and in-vitro release properties. Journal of Controlled Release 1998, 56 (1-3), 105-115.
31. Garcia del Barrio, G.; Novo, F.J.; Irache, J.M. Loading of plasmid DNA into PLGA microparticles using TROMS (Total Recirculation One-Machine System): evaluation of its integrity and controlled release properties. Journal of Controlled Release 2003, 86 (1), 123-30.
32. Santini, B.; Antonelli, M.; Battistini, A.; Bertasi, S.; Collura, M.; Esposito, I.; Di Febbraro, L.; Ferrari, R.; Ferrero, L.; Iapichino, L.; Lucidi, V.; Manca, A.; Pisconti, C.L.; Pisi, G.; Raia, V.; Romano, L.; Rosati, P.; Grazioli, I.; Melzi, G. Comparison of two enteric coated microsphere preparations in the treatment of pancreatic exocrine insufficiency caused by cystic fibrosis. Digestive and Liver Disease 2000, 32 (5), 406-411.
33. Elliott, R.B.; Escobar, L.C.; Lees, H.R.; Akroyd, R.M.; Reilly, H.C. A comparison of two pancreatin microsphere preparations in cystic fibrosis. The New Zealand Medical Journal 1992, 105 (930), 107-108.
34. Liu, H.W.; Ofosu, F.A.; Chang, P.L. Expression of human growth factor IX by microencapsulated recombinant fibroblasts. Human Gene Therapy 1993, 4 (3), 291-301.

35. Dubey, R.; Shami, T.C.; Bhasker K.U. Microencapsulation Technology and Applications. Defence Science Journal 2009, 59 (1), 82-95.
36. Gharsallaoui, A.; Roudaut, G.; Chambin, O.; Voilley, A.; Saurel, R. Applications of spray-drying in microencapsulation of food ingredients: an overview. Food Research International 2007, 40 (9), 1107-1121.
37. Klein, M.; Aserin, A.; Svitov, I.; Garti, N. Enhanced stabilization of cloudy emul- sions with gum Arabic and whey protein isolate. Colloids and Surfaces B: Biointerfaces 2010, 77, 75–81.
38. Guzey, D.; Kim, H.J.; McClements, D.J. Factors influencing the production of o/w emulsions stabilized by β-lactoglobulin-pectine membranes. Food Hydrocolloids 2004, 18, 967–975.
39. Gu, Y.S.; Regnier, L.; McClements, D.J. Influence of environmental stress on stability of oil-in-water emulsions containing droplets stabilized by β-lactoglobulin- κ-carrageenan membranes. Journal of Colloids and Interface Science 2005, 286, 551-558.
40. Moschakis, T.; Murray, B.S.; Biliaderis, C.G. Modifications in stability and structure of whey protein-coated o/w emulsions by interfacting chitosan and gum Arabic mixed dispersions. Food Hydrocolloids 2010, 24, 8-17
41. Hemar, Y.; Tamehana, M.; Munto, P. A.; Singh, H. Viscosity, microstructure and phase behaviour of aqueous mixtures of commercial milk protein products and xanthan gum. Food Hydrocolloids 2001, 15, 565-574.
42. Coultate, T.P. Alimentos: a química de seus componentes. Artmed: Porto Alegre, Brazil, 2004.
43. Idris, O.H.M.; Williams, P.A.; Phillips, G.O. Characterisation of gum from Acacia senegal trees of different age and location using multidetection gel permeation chromatography. Food Hydrocolloids 1998, 12, 379-388.
44. Connolly, S.; Fenyo, J.C.; Vandevelde, M.C. Effect of a proteinase on the macromolecular distribution of Acacia senegal gum. Carbohydrate Polymers 1988, 8, 23-32.
45. Islam, A.M.; Phillips, G.O.; Sljivo, A.; Snowden, M.J.; Williams, P.A. A review of recent developments on the regulatory, structural and functional aspects of gum arabic. Food Hydrocolloids 1997, 11, 493-505.
46. Renard, D.; Robert, P.; Lavenant, L,; Melcin, D.; Popineau, Y; Gueguen, J.; Duclairoir, C., Nakache, E.; Sanchez, C.; Schmitt, C. Biopolymeric colloidal carriers for encapsulation or controlled release applications. International Journal of Pharmaceutics 2002, 242, 163-166.
47. Kenyon, M.M. Modified starch, maltodextrin, and corn syrup solids as wall materials for food encapsulation. In Encapsulation and Controlled Release of Food Ingredients; Risch, S.J.; Reineccius, G.A., Eds.; American Chemical Society: Washington, DC, 1995; 43-50.

48. Schmitt, C.; Sanchez, C.; Thomas, F.; Hardy, J. Complex coacervation between beta-lactoglobulin and acacia gum in aqueous medium. Food Hydrocolloids 1999, 13, 483-496.
49. Gerritsen, J.; Crum, F. Getting the best from beta-carotene. International Food Ingredients 2002, 3, 40-41.
50. Junyaprasert, V.B.; Mitrevej, A.; Sinchaipanid, N.; Broome, P.; Wurster, D.E. Effect of process variables on the microencapsulation of vitamin A palmitate by gelatin-acacia coacervation. Drug Development and Industrial Pharmacy 2001, 27 (6), 561-566.
51. Trindade, M.A.; Grosso, C.R.F. The stability of ascorbic acid microencapsulated in granules of rice starch and in gum. Arabic. Journal of Microencapsulation 2000, 17 (2), 169-176.
52. Nixon, J.R; Agyilirah, G.A. The influence of colloidal proportions on the release of phenobarbitone from microcapsules. International Journal of Pharmaceutics 1980, 6, 271-283.
53. Xing, F.; Cheng, G.; Yang, B.; Ma, L. Microencapsulation of capsaicin by the complex coacervation of gelatin, acacia and tannins. Journal of Applied Polymer Science 2004, 91 (4), 2669-2675.
54. Takka, S.; Acarturk, F., Calcium alginate microparticles for oral administration: I: effect of sodium alginate type on drug release and drug entrapment efficiency. Journal of Microencapsulation 1999, 16, 275-290.
55. Cook, M.T.; Tzortzis, G.; Charalampopoulos, D.; Khutoryanskiy, V.V. Microencapsulation of probiotics for gastrointestinal delivery. Journal of Controlled Release 2012, 162, 56-67.
56. Rajaonarivony, M.; Vauthier, C.; Couarraze, G.; Puisieux, F.; Couvreur, P. Development of a new drug carrier made from alginate. Journal of Pharmaceutical Sciences 1993, 82, 912-917.
57. Chandramouli, V.; Kailasapathy, K.; Peiris, P.; Jones, M. An improved method of mcroencapsulation and its evaluation to protect Lactobacillus spp. in simulated astric conditions. Journal of Microbiological Methods 2004, 56, 27-35.
58. Sandoval-Castilla, O.; Lobato-Calleros, C.; García-Galindo, H.S.; Alvarez-Ramírez, J.; Vernon-Carter, E.J. Textural properties of alginate-pectin beads and survivability of entrapped Lb. casei in simulated gastrointestinal conditions and in yoghurt. Food Research International 2010, 43, 111-117.
59. Pandey, R.; Khuller, G.K. Alginate as a Drug Delivery Carrier. In Handbook of carbohydrate engineering; Yarema, K. J., Ed.; CRC Press: Boca Raton, 2005.
60. Joseph, I.; Venkataram, S. Indomethacin sustained release from alginate-gelatin or pectin-gelatin coacervates. International Journal of Pharmaceutics 1995, 126, 161-168.

61. Gonzalez-Rodriguez, M.L.; Holgado, M.A.; Sanchez-Lafuente, C.; Rabasco, A.M.; Fini, A. Alginate-chitosan particulate systems for sodium diclofenac release. International Journal of Pharmaceutics 2002, 232, 225-234.

62. Chickering, D.E.; Jacob, J.S.; Desai, T.A. Harrison, M.; Harris, W.P.; Morrell, C.N.; Chaturvedi, P.; Mathiowitz, E. Bioadhesive microspheres: An *in vivo* transit and bioavailability study of drug loaded alginate and poly (fumaric-co-sebacic anhydride) microspheres. Journal of Control Release, 1997, 48, 35-46.

63. Lannuccelli, V.; Coppi, G.; Cameroni, R. Biodegradable intraoperative system for bone infection treatment. I. The drug/polymer interaction. International Journal of Pharmaceutics 1996, 143, 195-201.

64. Cui, J.-H.; Goh, J.-S.; Park, S.-Y.; Kim, P.-H.; Lee, B.-J. Preparation and physical characterization of alginate microparticles using air atomization method. Drug Development and Industrial Pharmacy 2001, 27, 309-319.

65. Whitehead, L.; Collett, J.H.; Fell, J.T. Amoxycillin release from a floating dosage form based on alginates. International Journal of Pharmaceutics 2000, 210, 45-49.

66. Lucinda-Silva, R.M.; Evangelista, R.C. Microspheres of alginate-chitosan con- taining isoniazid. Journal of Microencapsulation 2003, 20, 145-152.

67. George, M.; Abraham, T. E. Polyionic hydrocolloids for the intestinal delivery of protein drugs: Alginate and chitosan - a review. Journal of Controlled Release 2006, 114, 1-14.

68. Tischer, P.C.S.F.; Noseda, M.D.; Freitas, R.A.; Sierakowski, M.R.; Duarte, M.E.R. Effects of iota-carrageenan on the rheological properties of starches. Carbohydrate Polymers 2006, 65, 49-57.

69. Osório, S. M. L. Novel polymeric systems based on natural materials: development and biological performance. Universidade do Minho. 2007

70. Bartkowiak, A.; D. Hunkeler. Carrageenan-oligochitosan microcapsules: optimization of the formation process. Colloids and Surfaces B:Biointerfaces 2001, 21 (4), 285-298.

71. Audet, P.; Paquin, C.; Lacroix, C. Immobilized growing lactic acid bacteria with k-carrageenan-locust bean gum gel. Applied Microbiology and Biotechnology 1988, 29, 11-18.

72. Ouellette, V.; Chevalier, P.; Lacroix, C. Continuous fermentation of a supplemented milk with immobilized Bifidobacterium infantis. Biotechnology Techniques 1994, 8, 45-50.

73. Doleyres, Y.; Paquin, C.; Leroy M.; Lacroix, C. Bifidobacterium longum atcc 15707 cell production during free- and immobilized-cell cultures in MRS-whey permeate medium. Applied Microbiology and Biotechnology 2002 60 (1-2), 168-173.

74. Suzuki, S.; Lim, JK. Microencapsulation with carrageenan-locust bean gum mixture in a multiphase emulsification technique for sustained drug release. Journal of Microencapsulation 1994, 11 (2), 197-203.
75. Sipahigil, O.; Dortunç, B. Preparation and *in vitro* evaluation of verapamil HCl and ibuprofen containing carrageenan beads. International Journal of Pharmaceutics 2001, 228 (1–2), 119-128
76. Tomida, H.; Nakamura, C.; Kiryu, S. A novel method for the preparation of controlled release theophylline capsules coated with a polyelecrolyte complex of kappa-carrageenan and chitosan. Chemical and Pharmaceutical Bulletin 1994, 42, 979-981.
77. Garcıa, A.M.; Ghaly, E.S. Preliminary spherical agglomerates of water soluble drug using natural polymer and cross-linking technique. Journal of Controled Release 1996, 40, 179-186.
78. Adzahan, N. M. Modification on wheat, sago and tapioca starches by irradiation and its effect on the physical properties of fish cracker (keropok). Food Technology. Selangor, University of Putra Malaysia. Master of Science 2002, 222.
79. Miyazaki, M.R.; Hung, P.V.; Maeda, T.; Morita, N. Recent advances in application of modified starches for breadmaking. Trends in Food Science and Technology 2006, 17, 591-599.
80. Uddin, M.S.; Hawlader, M.N.A.; Zhu, H.J. Microencapsulation of ascorbic acid: effect of process variables on product characteristics. Journal of Microencapsulation 2001, 18 (2), 199-209.
81. Dib Taxi, C.M.A.; Menezes, H.C.D.E.; Santos, A.B.; Grosso, C.R.F. Study of the microencapsulation of camu-camu (Myciaria dubia) juice. Journal of Microencapsulation 2003, 20 (4), 443-448.
82. Pariot, N., Edwwards-Lévy, F.; Andry, M-C., Lévy, M.-C. Cross-linked β-cyclodextrin microcapsules. II. Retarding effect on drug release through semi-permeable membrane. International Journal of Pharmaceutics 2002, 232, 175-181.
83. Larionova, N.V.; Ponchel, G.; Duchêne, D.; Larionova, N.I. Biodegradable cross-linked starch:protein microcapsules containing proteinase inhibitor for oral protein administration. International Journal of Pharmaceutics 1999, 189, 171-178.
84. Bulea, M.V.; Singhala, R.S.; Kennedy, J.F. Microencapsulation of ubiquinone-10 in carbohydrate matrices for improved stability. Carbohydrate Polymers 2010, 82, 1290-1296.
85. Pegg, R.B.; Shahidi, F. Encapsulation, Stabilization, and Controlled Release of Food Ingredients and Bioactives. In Handbook of Food Preservation; Rahman, M.S., Ed.; CRC Press: Boca Raton, 2007; 509-568.

86. Peniche, C.; Arguelles-Monal, W.; Peniche, H.; Acosta, N. Chitosan: An Attractive Biocompatible Polymer for Microencapsulation. Macromolecular Bioscience 2003, 3, 511-520.

87. Silva, C.F. Micropartículas de quitosana com didanosina e sua formulação em grânulos mucoadesivos, 2006. Universidade Estadual de Campinas, Faculdade de Engenharia Química: Campinas, Brazil.

88. Nisperos-Carriedo, M. O. Edible coatings and films based on polysaccharides. In Edible Coatings and Films to Improve Food Quality; Krochta, J. M.; Baldwin, E. A.; Nisperos-Carriedo, M. O., Eds.; Technomic Publishing Co. Inc.: Lancaster, 1994; 305.

89. http://www.lsbu.ac.uk/water/hycmc.html. (accessed December 2012)

90. Ahmad, M.; Madni, A.; Usman, M.; Munir, A.; Akhtar, N.; Khan, H.M.S. Pharmaceutical Microencapsulation Technology for Development of Controlled Release Drug Delivery systems. World Academy of Science, Engineering and Technology 2011, 51, 384-387.

91. Tiyaboonchai, W.; Ritthidej, G.C. Development of Indomethacin Sustained Release Microcapsules Using Chitosan-Carboxymethylcellulose Complex Coacervation. Songklanakarin Journal of Science and Technology 2003, 25, 245-254.

92. Kamal, A.H.M.; Ahmedl, M.; Wahed, M.I.I.; Amran, S.; Shaheenl, S.; Rashid, M. Development of Indomethacin Sustained Release Microcapsules using Ethyl Cellulose and Hydroxy Propyl Methyl Cellulose Phthalate by O/W Emulsification. Dhaka University Journal of Pharmaceutical Sciences 2008, 7 (1), 83-88.

93. http://www.nstedb.com/success/iedc/encapsulation.pdf (accessed December 2012)

94. Thibault, J.F.; Renard, C.M.G.C.; Guillon, F. Sugar beet fiber – production, composition, physicochemical properties, physiological effects, safety, and food application. In Handbook of Dietary Fiber; Cho, S.S.; Dreher, M.L., Eds; Marcel Dekker: New York, 2001, 553-582.

95. Ron, N. b-Lactoglobulin as a nano-capsular vehicle for hydrophobic nutraceuticals. Israel Institute of Technology, Haifa, Israel. 2007.

96. Craigic, J.S. Cell walls. In Biology of the red algae; Cole, K.M.; Sheath, R.G. Eds.; Cambridgc University Press: New York, 1990, 221-257.

8 Polysaccharides used in Nanoparticle based Drug Delivery Formulations

Sreeranjini Pulakkat and Krishna Radhakrishnan

Department of Materials Engineering, Indian Institute of Science, Bangalore, Karnataka, India-560 012.

8.1 Introduction

Nanoscience and nanotechnologies is a dynamic and fast developing field of research that can be regarded as genuinely interdisciplinary. An understanding of matter and processes at the nanoscale is relevant to all scientific disciplines, from chemistry and physics to biology, engineering and medicine. Nanoscience is defined as the study of phenomena and manipulation of materials at atomic, molecular and macromolecular scales, where properties differ significantly from those at a larger scale. Whereas, the design, characterization, production and application of structures, devices and systems by controlling shape and size at nanometre scale is usually termed as nanotechnology[1]. The inherent properties of nanomaterials were initially studied for their physical, mechanical, electrical, magnetic and chemical applications, but recent attention has been focussed towards their biological and pharmaceutical applications. Tremendous developments in the field of medical nanotechnology have shaped the concept of nanomedicine which attempts to provide personalized medicine by repairing or killing specific cells. Nanomedicine, as referred by the National Institutes of Health, is the application of nanotechnology for treatment, diagnosis, monitoring, and control of biological systems[2,3].

Nanoparticles are the most important tool employed in several nanomedical applications like gene therapy, imaging and novel drug discovery and drug delivery. Nanoparticles are defined as particulate dispersions or solid particles with a size in the range of 10-1000 nm. Depending upon the method of preparation, nanoparticles, nanospheres or nanocapsules can be obtained. The growing interest in the use of nanoparticles for drug delivery applications can be attributed to the outstanding advantages they offer when compared to large size materials.

Some of the limitations of conventional drug delivery include poor bioavailability, *in vivo* stability, solubility, intestinal absorption, sustained and targeted delivery to site of action, therapeutic effectiveness, generalized side effects, and plasma fluctuations of drugs. The size, geometrical shape and surface characteristics of nanoparticles can be controlled to achieve different properties and release characteristics of the drug that is dissolved, entrapped, encapsulated or attached to the matrix for the best delivery[4,5]. Nanoparticles having size less than 200 nm are usually pursued as drug delivery vehicles as that is comparable with the width of the microcapillaries[6]. Nanoparticles have the ability to protect drugs from degradation in the gastrointestinal tract, deliver poorly water soluble drugs, increase oral bioavailability of drugs due to their specialized uptake mechanisms such as absorptive endocytosis and are able to remain in the blood circulation for a longer time. They are able to bypass the liver, thereby preventing the first pass metabolism and can overcome various biological barriers. The small size of the nanoparticles enable passive targeting due to the enhanced permeability and retention (EPR) effect of the tumor vasculature. Additional functionalities like ligands, aptamers and small peptides, can be easily attached to nanoparticles that allows for molecular recognition of the target tissue or for active or triggered release of the payload at the disease site[7]. Nanoparticles can be tuned to show controlled release properties due to the biodegradability, pH ion and/or temperature sensibility of materials and hence improve the utility of drugs and reduce toxic side effects. The nanoparticle system can be used for various routes of administration including oral, nasal, parenteral, intra-ocular etc. Multi functional nanoparticles enable combination therapy in which delivery of two or more drugs is possible along with real time read on their *in vivo* efficacy. There have also been attempts at synthesizing theranostic nanoparticles that can simultaneously deliver imaging and therapeutic agents to specific sites or organs, enabling detection and treatment of disease in a single procedure[8,9]. A few limitations like particle-particle aggregation, difficulty in physical handling of nanoparticles in liquid and dry forms, burst release etc., need to be overcomed before nanoparticles can be used clinically or made commercially available. Also, the safety of nano-drug delivery systems in humans are of great concern these days as there has been reports that smaller nanoparticles show increased toxicity due to their increased surface area[10]. Hence these days, much research effort has been devoted to the use of natural polymers in drug delivery systems that always show low/non toxicity, biodegradability, low immunogenicity and good biocompatibility[11,12].

Natural polysaccharides, due to their outstanding merits, are the most popular polymeric materials used in the development of nano-sized drug delivery systems. Polysaccharides are the most abundant macromolecules in the biosphere and are made up of repeated mono or disaccharides joined by glycosidic bonds. They are sometimes also called glycans. For polysaccharides which contain a substantial proportion of amino sugar residues the term glycosaminoglycan is a common one. Polysaccharides which consist of only one kind of monosaccharide are called homopolysaccharides; while those which are built up of two or more different monomeric units are named heteropolysaccharides. Homo-as well as heteropolysaccharides can be linear or branched. They have a general formula of $C_x(H_2O)y$ where x is usually a large number between 200 and 2500. The repeating units in the polymer backbone are often six-carbon monosaccharides, hence the general formula can also be represented as $(C_6H_{10}O_5)n$ where $40 \leq n \leq 3000$. They are often one of the main structural elements of plants and animals exoskeleton or have a key role in the plant energy storage. In nature, polysaccharides have various resources from algal origin (e.g., alginate), plant origin (e.g., pectin, guar gum), microbial origin (e.g., dextran, xanthan gum), and animal origin (chitosan, chondroitin)[13].

Owing to their outstanding virtues, polysaccharides have received the majority of attention in the field of nanoparticle drug delivery systems. They are biocompatible and as a result are non-toxic in humans. They tend to be internalized and degraded rapidly, thus enabling a moderate intracellular release of the drug or gene. Due to the presence of various reactive groups in their backbone, polysaccharides can be easily modified chemically and biochemically to create derivatives with determined/ tailored properties. The presence of functional groups also allows for conjugation of therapeutic drugs to the main chain through functional linkers[14,15]. The presence of hydrophilic groups in their structure such as hydroxyl, carboxyl and amino groups, enhance bioadhesion with biological tissues like epithelia and mucous membranes forming non-covalent bonds, which helps to enhance *in vivo* residence times in the gastrointestinal tract and consequently the amount of absorbable drug. In addition, positively charged polysaccharides are also capable of opening the tight junctions between epithelial cells, thereby increasing the paracellular permeability of hydrophilic drugs across the mucosal epithelia. Owing to these characteristics, mucoadhesive polysaccharides have been widely used to deliver vaccines and hydrophilic macromolecular therapeutics through pulmonary or nasal routes[16]. Some polysaccharides possess the innate ability to recognize specific receptors that are over-expressed on the surfaces of diseased cells and allow the design of carriers that can selectively deliver active agents through receptor mediated endocytosis. They also allow for non-specific protein adsorption by providing neutral coating with low surface energy[17,18]. Hydrophobic moieties can be attached to polysaccharides to produce amphiphilic derivatives that are able to self-assemble in biological fluids and act as carriers for the delivery of hydrophobic drugs. Another main advantage of polysaccharides is their availability in natural resources and low cost in their processing, which make them very accessible materials to be used as drug carriers. Most of the polysaccharides are hydrophilic in nature enabling simple and mild preparation methods[19,20]. All these merits endow polysaccharides as promising drug delivery carriers that are suitable for a broad category of drugs including macromolecules and labile drugs and the issues of safety toxicity and availability are greatly simplified. In recent years, a large number of studies have been conducted on polysaccharides and their derivatives for their potential application as nanoparticle drug delivery systems.

8.2 Polysaccharide-based Nanoparticles Fabricated by Crosslinking

8.2.1 Covalently Crosslinked Polysaccharide Nanoparticles

This class of nanoparticles are fabricated by covalently linking polysaccharide chains to each other thereby incorporating novel functionalities into the polysaccharide. The chemical links are formed by using a chemical crosslinker which connects two functional groups present on the polysaccharide chains. The crosslinkers are generally homo or hetero-bifunctional molecules containing reactive end groups which are capable of forming chemical bonds with the functional groups present on the polysaccharide molecules. The crosslinker utilizes the reactive functional groups such as hydroxyls, amines and carboxyls present on the backbones of the polysaccharides to form stable covalent bonds. The result is the formation of intra-molecular crosslinks (within the same polysaccharide molecules) or interchain crosslinks (between two different polysaccharide molecules).

The dialdehyde homo-bifunctional crosslinker glutaraldehyde was initially the most commonly used crosslinker for preparation of crosslinked polysaccharide nanoparticles[21]. The aldehyde groups of glutaraldehyde react with amine groups present on polysaccharides thereby forming stable covalent crosslinks. Mitra et al. reported the preparation of chitosan nanoparticles by glutaraldehyde crosslinking. They employed a reverse microemulsion technique to restrict the particle size to 100 ± 10 nm. These nanoparticles were capable of carrying the anticancer drug doxorubicin to macrophage tumor cells implanted in Balb/c mice[21]. Gluteraldehyde was later replaced by other crosslinkers such as carbodiimides which allowed the crosslinking reactions to be carried out in mild and aqueous reaction conditions. Carbodiimides are "zero length" crosslinkers which assist the formation of amide linkages by condensation of amino group and carboxylic acid groups. EDC [or EDAC; l-ethyl-3-(3-dimethylaminopropyl) carbodiimide hydrochloride], l-cyclohexyl-3-(2-morpholinoethyl) carbodiimide (CMC), dicyclohexyl carbodiimide (DCC) are some of the carbodiimides used in polysaccharide crosslinking. Recently natural di- or tri carboxylic acids such as, malic acid, tartaric acid, succinic acid and citric acid have been used as biocompatible chemical crosslinkers[22].

Several groups have utilized an emulsion crosslinking method for the production of polysaccharide nanoparticles wherein water-in-oil emulsion is employed to carry out the crosslinking reaction. The water droplets finely dispersed in the oil media act as nanosized reaction vessels which determined the size of the nanoparticles formed. Here the water-in-oil emulsion is formed by dispersing the aqueous polysaccharide solution in an oil phase. This emulsion is further stabilized by a surfactant and later crosslinked using an appropriate crosslinking agent. The particles can be isolated by repeated washing and filtering steps[22,23]. Emulsion crosslinking often produces large particles (> 200 nm) and with a broad size range. This can be overcome by using a reverse micellar process which produces ultrafine nanoparticles (1−10 nm) with a narrow size distribution[24]. Reverse micelles are liquid mixtures consisting of water, oil and a surfactant. The reverse micellar droplets formed by dissolving the surfactant in an organic solvent, are extremely stable and form ultrafine nanoreactors for the crosslinking reaction. The polysaccharide solution along with the drug to be loaded is added to this while maintaining constant vortexing. This is followed by the addition of the crosslinking molecule resulting in the formation of nanoparticles. The organic solvent is evaporated and the surfactant is removed by salt precipitation to obtain the drug loaded nanoparticles[25].

8.2.2 Ionically Crosslinked Polysaccharide Nanoparticles

Another major fabrication strategy is ionic crosslinking whereby polycationic or polyanionic. Polysaccharides can be crosslinked using ionic crosslinkers such as charged polysaccharides, low molecular weight polyanions and polycations. This offers several advantages such as mild reaction conditions, simple preparation steps and biocompatible by-products. This technique also provides the advantage of easy scale up. Since this technique can be carried out at room temperature without organic solvents, it can be employed for encapsulation and delivery of sensitive drug molecules such as protein and DNA based drugs[26-35]. In fact many groups have employed charged drug molecules as the ionic crosslinker to fabricate polysaccharide nanoparticles. For example Pan et al. synthesized chitosan-TPP-insulin nanoparticles by utilizing the electrostatic interaction between the amine groups of chitosan and the acidic insulin groups[35]. These nanoparticles were proven to enhance the intestinal absorption of insulin and thereby better pharmacological bioavailability in diabetic rats.

Cationic polysaccharides are often ionically crosslinked with the negatively charged DNA as non-viral carriers for gene therapy applications. Howard et al. reported the fabrication of chitosan siRNA nanoparticle delivery system for treatment of systemic and mucosal diseases. Chitosan dissolved in acetic acid was mixed with siRNA agent that had a pre-optimized nucleotide length. This lead to the development of ionic crosslinks between chitosan polymeric chains culminating in the formation of chitosan siRNA nanoparticles. These particles mediated the knock down of the endogenous enhanced green fluorescent protein in the *in vitro* studies carried out using H1299 human lung carcinoma cells and murine peritoneal macrophages. The *in vivo* studies using these nanoparticles showed effective RNA interference in bronchiole epithelial cells of transgenic EGFP mice[30]. Ionic crosslinking has often been used in combination with emulsification processes such as water-in-oil emulsions wherein the ionic gelation of the nanoparticles occur inside the nanosized water droplets which restricts the size of the final nanoparticles being formed[36,37].

Chitosan and chitosan derivatives which have been reported to show biodegradability, biocompatibility, hemostatic, antimicrobial and anticholesteremic properties are the most exploited polysaccharides for the preparation of ionically crosslinked nanoparticles. These polysaccharides are co-incubated with ionic crosslinkers of opposite charges to obtain nanoparticles with varying properties. For example, the high number of cationic amine groups attached to the glucosamine units present on acetylated chitosan backbone form intrachain and interchain crosslinks with anionic crosslinkers such as tripolyphosphate (TPP), (DNA) etc., forming various types of nanoparticles[29,31,32,38]. Of these, chitosan nanoparticles in combination with TPP have been widely utilized for drug delivery[21]. Chitosan- TPP nanoparticles are prepared by mixing chitosan dissolved in acidic solution (pH 4-6) with an alkaline solution of TPP. The phosphate group present in the TPP molecules create intermolecular as well as intramolecular linkages with amino groups present on chitosan. The characteristics of so formed nanoparticles can be carefully designed by varying the reaction parameters such as concentrations of chitosan and TPP, molecular weight of chitosan employed, purity of the acid salt etc[39]. Although initially these nanoparticles were fabricated from pure chitosan and TPP, later studies demonstrated addition of other polymers and macromolecules in the reaction mixtures to create hybrid nanoparticles. Depending on the characteristics of these polymers or macromolecules, additional properties such as enhanced interactions with deliverables like protein drugs, targetability of the drug delivery system, control over the net charge of the nanoparticles etc., are achieved[38].

Negatively charged polysaccharides such as alginate can form nanoparticles via inter-and intramolecular crosslinking with multivalent cationic ions such as calcium chloride resulting in precipitation. The content of guluronic acid and mannuronate blocks in the polysaccharide decides the strength of the nanogels formed. A higher content of guluronic acid residues gives greater strength compared to mannuronate rich alginates. This is due to the stronger affinity exhibited by the guluronic acid residues to divalent ions. Several deliverables such as insulin, antimicrobial drugs, genes etc.,[37,41-43] have been incorporated into alginate based nanoparticles. Initially alginate nanoparticles fabrication was carried out by adding low concentration of calcium ions to dilute solutions of alginate. This leads to the formation of clusters of calcium alginate gels[43]. This was later replaced by an emulsification method where the gelation process was carried out in droplets of required sizes thereby restricting the sizes of the nanoparticles formed[37]. Several reports have emerged demonstrating encapsulation and targeted delivery of drugs using hyaluronic acid based nanoparticles. Hyaluronic acid is a polyanionic biopolymer that has the ability to target $CD44^+$ receptors present on some cancer cells[45-47]. Nanoparticles fabricated from hyaluronic acid has the ability to specifically accumulate in tumor sites by utilizing the enhanced permeability and retention effect (EPR) and receptor mediated endocytosis[48,49].

8.3 Self-Assembling Polysaccharide-based Nanoparticles

8.3.1 Self-assembly of Hydrophobically Modified Polysaccharides

Amphiphilic polysaccharide derivatives are synthesized by grafting hydrophobic moieties onto a hydrophilic polysaccharide. In aqueous solutions they tend to self-assemble into nanoparticles with multiple hydrophobic inner cores covered by a hydrophilic polysaccharide shell. This self-assembly process is driven by intra- or intermolecular interactions between hydrophobic moieties, primarily to minimize interfacial free energy. The hydrophilic shell serves as a stabilizing interface between the inner hydrophobic core and the external aqueous environment. These self-assembled polysaccharide nanoparticles possess unique characteristics such as unusual rheological feature, small hydrodynamic radius, core-shell structure, prolonged circulation and thermodynamic stability. The size, surface charge, loading efficiency, stability and biodistribution of these nanoparticles can be altered to suit for different applications. For example, the length of the hydrophobic moiety and the length of the polymer can be altered to control the size of the nanoparticles. Further, the surface charge, which affects particle serum stability and cellular uptake, can be altered by controlling the degree of substitution, the length or nature of the hydrophobic moiety. However, it is necessary to avoid excess hydrophobic modification as this may result in the loss of the physicochemical and biological characteristics of the parent polysaccharides or precipitation of the polymeric amphiphiles under aqueous conditions[49].

The self-assembled polysaccharide nanoparticles have been recognized as promising drug carriers owing to their hydrophobic cores that act as reservoirs for the delivery of several poorly water-soluble drugs. Several polysaccharides such as dextran, chitosan, glycol chitosan, hyaluronic acid (HA), heparin, and pullulan have been hydrophobically modified to prepare self-assembled nanoparticles. The hydrophobic moieties used range from linear lipophilic molecules, cyclic hydrophobic molecules, hydrophobic drugs, oligomers to polymers from polyacrylate family etc. In the past few years, several studies have been carried out to investigate the synthesis and the application of self-assembled polysaccharide nanoparticles as drug delivery systems.

8.3.1.1 Linear Hydrophobic Molecules

Polyethyleneglycol (PEG) has been employed extensively in biomedical research as a soluble polymeric modifier in organic synthesis owing to its outstanding physicochemical and biological properties such as high hydrophilicity, solubility, biocompatibility, biodegradability, ease of chemical modification and absence of antigenicity and immunogenicity. It enables prevention of bacterial surface growth, decrease of plasma protein binding and erythrocyte aggregation, and prevention of recognition by the immune system. Recently, PEG grafted chitosan and chitosan derivatives have been widely studied by many researchers[50]. Yoksan et al. grafted PEG-methyl ether onto N-Phthaloyl chitosan chains and obtained stable sphere-like nanoparticles with size as small as 80-100 nm by simply adjusting the hydrophobicity/hydrophilicity of the chitosan chain[51]. Whereas, Opanasopit et al. used N-phthaloyl chitosan-grafted PEG-methyl ether, to form spherical core–shell micelle-like nanoparticles exhibiting sizes in the range of 100-250 nm[52]. Methoxy PEG-grafted-chitosan conjugates was used by Jeong et al.[53] and Yang et al.[54] to develop polymeric micelles and monodisperse self-aggregated nanoparticles respectively for drug delivery applications.

Poly(ε-caprolactone) (PCL), a biodegradable polyester with excellent mechanical strength, biocompatibility and nontoxicity, has been combined with polysaccharides to produce amphiphilic copolymer drug delivery systems. Yu et al. synthesized biodegradable amphiphilic PCL-graft-chitosan copolymers which could form spherical or elliptic nanoparticles in water[55]. The hydroxyl groups of dextran with the carboxylic function have been coupled with preformed PCL blocks to form nanoparticles of less than 200 nm. Further, bovine serum albumin and lectin were incorporated within and onto the surface of the nanoparticles for facilitating targeted oral administration[56,57].

Pluronic tri-block copolymers self-assemble to form a spherical micellar structure above the lower critical solution temperature by hydrophobic interaction of the poly(propylene oxide) middle block in the structure. Hence they have been used to prepare thermoresponsive composite nanocapsules along with heparin, chitosan etc. Core–shell Pluronic-chitosan nanocapsules designed to encapsulate small molecules for temperature-controlled release and intracellular delivery was developed by Zhang et al. and the thermal responsiveness of the nanocapsules in size and wall-permeability was studied [58]. Chitosan/pluronic hydrogels have also been used to encapsulate human dermal fibroblasts, deliver anticancer drugs, and as an injectable cell delivery carrier for cartilage regeneration[59,60,61]. Dai Hai Nguyen et al. prepared disulfide-crosslinked heparin-Pluronic nanogels encapsulated with RNase and characterized the *in vitro* release and cytotoxicity in the presence of glutathione[62]. Pluronic/heparin composite nanocapsules synthesized by Choi et al exhibited a 1000-fold volume transition and a reversible swelling and de-swelling behaviour with changes in temperature[63].

Modified polysaccharides synthesized using long-chain fatty acids such as hexanoic acid, decanoic acid, linoleic acid, linolenic acid, palmitic acid, stearic acid, and oleic acid have been used to prepare self-assembled nanoparticles by many researchers. Self-organized nanoparticles from decanoate β-cyclodextrin esters and hexanoate β-cyclodextrin esters biocatalyzed by thermolysin from native β-cyclodextrin using vinyl decanoate or vinyl hexanoate as acyl donors were synthesized by Choisnard et al.[64]. Chen et al. modified chitosan by coupling with linoleic acid through the EDC-mediated reaction to increase its amphipathicity for improved emulsification. The self-aggregated micelles thus formed, having size ranging from 200 to 600 nm, were used to encapsulate a lipid soluble model compound, retinal acetate, with 50% efficiency[65]. The same group modified chitosan with linolenic acid which resulted in self aggregates in which bovine serum albumin was loaded as a model drug[66]. These nanoparticles were also used to immobilize trypsin using glutaraldehyde as crosslinker. The kinetic constant value of trypsin immobilized on nanoparticle and the thermal stability was higher than that of pure trypsin which makes it more attractive in the application aspect[67]. In similar lines, Hu et al.[68] synthesized stearic acid grafted chitosan oligosaccharide by EDC-mediated coupling reaction and the shells were cross-linked by glutaraldehyde to increase the *in vivo* stability and controlled release of an anticancer drug paclitaxel. Jiang et al. prepared water-soluble Npalmitoyl chitosan micelles in water by coupling swollen chitosan with palmitic anhydride in dimethyl sulfoxide. Hydrophobic model drug ibuprofen was loaded in the micelles, the release of which strongly depended on pH and temperature[69]. Self-assembled nanoparticles based on oleoyl-chitosan developed by Zhang et al. showed negligible hemolysis rates and cytotoxicity. The loaded drug doxorubicin was rapidly and completely released from the nanoparticles at pH 3.8, whereas at pH 7.4 there was a sustained release after a burst release[70].

Dextran has been also employed to obtain nanoparticles by coupling lipoic acid to its structure[44]. Amylose conjugated linoleic acid complexes were synthesized to serve as molecular nanocapsules for the protection and the delivery of linoleic acid[71].

8.3.1.2 Cyclic Hydrophobic Moieties

Conjugating hydrophobic cholesterol to hydrophilic polysaccharides forms amphiphilic copolymer which may further form self-assembled nanoparticles in aqueous solution. Wang et al. synthesized cholesterol-modified chitosan conjugate with succinyl linkages and cholesterol-modified O-carboxymethyl chitosan which then formed monodisperse self-aggregated nanoparticles in aqueous media[72]. The Akiyoshi's group developed self-assembled cholesterol-bearing pullulan (CHP) nanoparticles which exhibited high colloidal stability, and no precipitation or dissociation of insulin incorporated within them. Further, the nanoparticles greatly improved the thermal stability of insulin, protected from enzymatic degradation and preserved its bioactivity *in vivo* after intravenous injection[73]. The authors have also demonstrated that these nanoparticles have chaperone-like activity wherein, thermally or chemically denatured proteins form complexes with CHP and upon addition of *β*-cyclodextrin, the refolded proteins are released[74]. In addition, they also prepared thermo-responsive nanoparticles by self-assembly of cholesterol-pullulan and poly(N-isopropylacrylamides)[75].

It is expected that the introduction of amphiphilic bile acids such as deoxycholic acid or 5β-cholanic acid into chitosan would induce self-association to form self-aggregates. Thus, Lee et al. covalently conjugated deoxycholic acid to chitosan via carbodiimide-mediated reaction to generate self-aggregated nanoparticles which were then used to physically entrap an anthracycline drug, adriamycin[76]. They have also utilised these chitosan self-aggregates for gene delivery applications[77]. Chae et al. chemically modified chitosan oligosaccharides with deoxycholic acid to form self-aggregated nanoparticles in the range of 200–240 nm which showed superior gene condensation, protection of condensed gene from endonuclease attack and high level of gene transfection efficiencies, even in the presence of serum[78].

However, chitosan-based self-aggregates being insoluble in biological solution (pH 7.4) are now replaced by water-soluble chitosan derivatives as novel drug carriers with increased stability and decreased cytotoxicity. Kim et al. prepared deoxycholic acid and 5β-cholanic acid modified glycol chitosan self-aggregates as new drug delivery systems. These nanoparticles were used to load Arg–Gly–Asp peptide that specifically binds to avß3 integrin expressed on endothelial cells in the angiogenic blood vessels[79-80]. In addition, the 5β -cholanic acid-glycol chitosan was also used to spontaneously form self-assembled nanoparticle with DNA having increased *in vitro* and *in vivo* transfection efficiencies[81]. The 5β-cholanic acid-glycol chitosan self-assembly nanoparticles were also used as carriers for sustained release of paclitaxel, an anticancer agent[82]. Recently, Cy5.5- labelled hydrophobically-modified glycol chitosan nanoparticles with prolonged circulation in the blood and excellent *in vivo* tumor specificity proved to be highly promising carriers for diagnosis of early stage cancers and drug delivery of various anticancer drugs[83]. N-acetyl histidine-conjugated glycol chitosan self-assembled nanoparticles developed by Park et al. also proved to be a promising carrier for intracytoplasmic delivery of drugs[84]. They also synthesized deoxycholic acid-heparin amphiphilic conjugates with different degree of substitution (DS) of deoxycholic acid, which provided monodispersed self-aggregates in water, with mean diameters (120–200 nm) decreasing with increasing DS[85]. Park's group has also prepared several amphiphilic hyaluronic acid (HA) derivatives by chemically conjugating HA with 5*β*-cholanic acid or hydrotropic oligomer which were able to form self-assembled nanoparticles under aqueous conditions[86,87]. Fluorescein isothiocyanate (FITC) and doxorubicin themselves are hydrophobic cyclic molecules, which form amphiphilic copolymers upon conjugation onto hydrophilic polysaccharides. Park et al. prepared hydrophobically modified glycol chitosans by chemical conjugation of FITC or doxorubicin to the backbone of glycol chitosan and evaluated the biodistribution of self-aggregates (300 nm in diameter) after systemic administered via the tail vein[88,89]. Cho et al. studied *in vivo* tumor targeting and radionuclide imaging with FITC-conjugated glycol chitosan nanoparticles in terms of mechanisms, key factors, and their implications[90]. Another derivative of chitosan, N-succinyl chitosan was successfully synthesized by Zhu et al., which self-assembled into well-dispersed and stable nanospheres in distilled water having 50–100 nm in diameter[91]. Na et al. have achieved active targeting by incorporating vitamin H along with pullulan acetate and the corresponding 100 nm sized self-assembled nanoparticles were loaded with adriamycin in order to improve their cancer-targeting activity and internalization[92].

8.3.1.3 Polyacrylate Family Molecules

Poly(methyl methacrylate) and poly(isobutyl cyanoacrylate) (PIBCA) are widely used hydrophobic biomaterials in the polyacrylate family that contain carboxylic ester groups in their structures. Passirani et al. prepared nanoparticles bearing heparin and dextran covalently bound to poly(methyl methacrylate) and evaluated their interactions with complement system. Both nanoparticles were weak activators of complement and possessed the "stealth" effect exhibiting long circulation time and reduced uptake by mononuclear phagocyte system *in vivo*[93,94]. Among the different polysaccharides, chitosan and dextran are good examples of hydrophilic molecules that have been used with PIBCA to obtain nanometric micelles. Bertholon et al. prepared PIBCA-chitosan, PIBCA-dextran and PIBCA-dextran sulfate core-shell nanoparticles by redox radical or anionic polymerization of IBCA in the presence of corresponding polysaccharides used[95]. Emulsion polymerization of IBCA in the presence of chitosan as a polymeric stabilizer at low pH yielded PIBCA-chitosan nanoparticles in which nimodipine was successfully incorporated as a model drug[96]. Radical emulsion polymerization of IBCA in the presence of various polysaccharides like dextran, dextran sulfate, heparin, chitosan, thiolated chitosan, hyaluronic acid, pectin etc. was utilised to obtain various nanoparticulate systems[97,98]. These nanoparticles were then further investigated for their calcium binding capacity, mucoadhesion mechanism and *in vitro* interactions of with blood proteins[99-100]. Chauvierre et al. also investigated the antithrombic activity of heparin in heparin-PIBCA copolymers which were further used to prepare nanoparticles that can carry haemoglobin and act as suitable tools in the treatment of thrombosis oxygen deprived pathologies[101].

8.3.2 Polysaccharide-based Polymersomes

Polymersomes are spherical vesicles made from amphiphilic block copolymers. Owing to their improved stability and capability to encapsulate both hydrophilic and hydrophobic drugs, polymersomes have emerged as candidates for drug delivery applications[102]. Most of the polymersomes have been prepared from synthetic block copolymers and hence are not recommended for clinical purposes. As a result, polysaccharide based polymersomes are gaining importance as biocompatible drug carriers. Lecommandoux et al. prepared polymersomes using amphiphilic dextran-block poly(γ-benzyl L-glutamate) (Dex-*b*-PBLG) and HA-*b*-PBLG block copolymers and used them for delivery of the anticancer drugs such as DOX[103,104]. Otsuka et al reported preparation of a thermoresponsive saccharide-containing block copolymer by coupling maltoheptaose to poly(N-isopropylacrylamide) via click chemistry to obtain polymersomes[105].

Another example of a polysaccharide polymersome involved a β-cyclodextrin head group coupled to polystyrene which also formed vesicles and was used to immobilize enzymes[106]. In polymersomes obtained using block copolymers, the biological activity of the polysaccharides is retained, because chemical conjugation is performed only at the chain end. However, this method hasn't gained widespread application because of the limited accessibility of the reducing-end groups and difficulty in selecting compatible solvents for the two blocks.

8.4 Polysaccharide Nanoparticles by Polyelectrolyte Complexation

Polyelectrolyte complexes (PECs) are formed by the intermolecular electrostatic interaction between oppositely charged polysaccharides. These materials possess interesting properties, like swelling or permeability that respond to external stimuli, such as the pH of the medium. Positively or negatively charged nanoparticles with a core/shell structure can be obtained depending on the nature of the polyelectrolyte used in excess. The excess polyelectrolyte not incorporated in the hydrophobic core is segregated in the outer shell imparting colloidal stability and charge to the nanoparticles. The size of the PEC nanoparticles can be varied by means of adjusting the molecular weight and charge density of component polymers. Mostly water-soluble and biocompatible polyelectrolytes are used to fabricate PEC nanoparticles for safety purposes. In this sense, chitosan is the widely used natural polycationic, polysaccharide that satisfies the needs of safety and solubility. There are many negative polymers complexed with chitosan to form PEC nanoparticles, which fall in the categories of polysaccharides, peptides, polyacrylic acid family and so on.

8.4.1 Negative Polysaccharides

Polyanionic polysaccharides such as HA, alginate, carboxymethyl cellulose, dextran sulfate, heparin etc., have been used with chitosan to form PEC. Cui et al.[107] prepared stable cationic nanoparticles containing plasmid DNA as a potential approach to genetic immunization by complexing carboxymethyl cellulose with chitosan. Chitosan/dextran sulfate nanoparticle delivery system whose physicochemical and release characteristics could be modulated by changing ratios of two ionic polymers was developed using a simple coacervation process[108]. Tiyaboonchai et al. used zinc sulfate as a crosslinking and hardening agent in a nanoparticulate delivery system for amphotericin B with chitosan and dextran sulphate[109]. In another study, vascular endothelial growth factor (VEGF)-loaded nanoparticles were prepared using dextran sulfate and chitosan[110]. Sarmento et al. synthesized nanoparticles by ionotropic complexation and coacervation between polyanions (dextran sulfate and alginate) and chitosan. Dextran sulfate/chitosan nanoparticle system provided highest insulin association efficiency and retention of insulin in gastric simulated conditions[111]. They also prepared insulin loaded nanoparticles by ionotropic pre-gelation of alginate with calcium chloride followed by complexation between alginate and chitosan[112]. Complexation of the cationic trimethyl chitosan and anionic cisplatin-alginate yielded nanoparticles having size in the range of 180 to 350 nm. The smaller nanoparticles with a low zeta potential were found to be more active than native cisplatin against human A549 lung cancer cells[113]. Li et al. prepared quaternized chitosan/alginate nanoparticles of size 200 nm for the oral delivery of protein[114]. Two different types of glucomannan (non-phosphorylated and phosphorylated) were used by Alonso-Sande et al. to form PEC by interaction with chitosan in the presence and absence of an ionic cross-linking agent, sodium tripolyphosphate[115]. Du et al. prepared carboxymethylkonjac-glucomannan/ chitosan nanoparticles with pH-responsive properties and ionic strength-sensitive properties under very mild conditions via polyelectrolyte complexation[116]. Liu et al. prepared Heparin/chitosan nanoparticles were also prepared by polyelectrolyte complexation in which bovine serum albumin was incorporated into as a model protein drug and the encapsulation efficiency along with *in vitro* release behaviour were investigated[117].

8.4.2 Negative Peptides

Apart from oppositely charged polysaccharides, PEC nanoparticles have also been prepared by complexation of polysaccharides with bio macromolecules such as proteins and nucleic acids. For example, PECs of chitosan derivatives and insulin were formed above the critical pH value of 6.0, in which both chitosan derivatives and insulin are ionized and bear the opposite charge[118]. Similarly, chitosan-DNA PEC nanoparticles possessing a high transfection efficiency have been prepared by mixing an optimum ratio of chitosan with plasmid DNA[119]. Nanocarriers developed by deMartimprey et al. had chitosan along with nucleic acids performing dual function of a structural component and therapeutic agent[120]. Apart from chitosan, polyelectrolyte complexes can be formed using other negatively charged polysaccharides like alginate and positively charged peptides like polylysine[121]. Another negatively charged polysaccharide HA was used to prepare nanocomplexes with a positively charged therapeutic protein for cancer and rheumatoid arthritis, tumor necrosis factor (TNF)-related apoptosis inducing ligand[122].

8.4.3 Others

A lot of researchers have investigated the formation of polyelectrolyte complexes between chitosan and other synthetic polymers like poly(acrylic acid) (PAA), poly-γ-glutamic acid, polyaspartic acid etc. The variation of size, stability and morphology of the nanoparticles with changes in molecular weight and concentration of chitosan and PAA, temperature and pH of the initial solutions has been thoroughly investigated[123,124]. Poly-γ-glutamic acid/chitosan nanoparticle system with enhanced intestinal paracellular transport was developed by Lee et al. using ionic-gelation method and further used for transdermal gene delivery[125]. Zheng et al. prepared anionic or cationic nanoparticles based on chitosan and polyaspartic acid sodium salt containing a hydrophilic antimetabolite drug, 5-fluorouracil[126]. The same group also prepared chitosan/glycyrrhetic acid nanoparticles and studied the glycyrrhetic acid encapsulation efficiency and *in vitro* release[127]. Polymethacrylic acid/chitosan/polyethylene glycol nanoparticles prepared by Sajeesh et al. exhibited good protein encapsulation efficiency and pH responsive release profile under *in vitro* conditions. Insulin and bovine serum albumin were incorporated into these nanoparticles as model proteins via diffusion filling method[128]. The development and exploitation of a lot of other biocompatible polycationic and polyanionic polysaccharides along with their water-soluble derivatives fuels the research in this field of polyelectrolyte nanoparticle systems.

8.5 Polysaccharide-based Hollow Nanospheres

Hollow polymer nanospheres or polymer nanocapsules are versatile nanostructures that have received enormous attention in drug delivery research owing to their ability to encapsulate large amounts of bioactive agents such as proteins, peptides, and nucleic acids. Their structural characteristics such as shape, size of the interior cavity, shell thickness, permeability, composition, and surface charge can be finely controlled to suit different applications. Layer-by-layer deposition of polyelectrolytes on a template core, self-assembly of block copolymers in selected solvents, and emulsion polymerization techniques are some of the established methods used to fabricate hollow nanospheres[129,130].

Layer by layer (LbL) method involves assembly of oppositely charged polyelectrolytes onto colloidal particles, followed by removal of the template cores. This method enables to obtain hollow nanospheres with well-controlled size and shape, finely tuned wall thickness, and variable wall compositions. A lot of polysaccharide polyelectrolyte combinations have been utilised so far to fabricate hollow multilayer nanocapsules or nanospheres. Most polysaccharides are negatively charged and are thus used as polyanion constituents, unless they are chemically modified to render them polycationic. Dextran, alginate, HA, carboxymethylcellulose, heparin, pectin, polygalacturinic acid etc., are some of the polyanionic polysaccharides that have been successfully used. The choice of a polycationic polysaccharide is very limited. In fact, chitosan is probably by far the most widely used polysaccharide in LbL assembly[131,132]. Single component polyelectrolyte capsules have also been fabricated, for example, Wang et al. fabricated chitosan hollow nanospheres using biodegradable poly(D,L-lactide) (PLA)-*b*-PEGblock copolymer nanoparticles as a template and by crosslinking the adsorbed chitosan using gluteraldehyde[133]. Some of the researchers have attempted template free methods to prepare hollow nanospheres. Hu et al. prepared chitosan-poly(acrylic acid) hollow nanospheres by polymerizing acrylic acid monomers in the presence of chitosan, followed by selective crosslinking of chitosan using glutaraldehyde. These nanospheres were further used to encapsulate an anticancer drug, doxorubicin[134]. In another template-free method for the preparation of crosslinked hollow nanospheres, Yin et al. utilised a single step reaction between azidobenzaldehyde and carboxymethyl chitosan in water at room temperature to fabricate nanospheres[135].

References

1. Nanoscience and nanotechnologies: opportunities and uncertainties | Royal Society." [Online]. Available: http://royalsociety.org/policy/publications/2004/nanoscience-nanotechnologies/. [Accessed: 14-May-2013].
2. R. Freitas, "Current Status of Nanomedicine and Medical Nanorobotics," J. Comput. Theor. Nanosci., vol. 2, no. 1, pp. 1-25, Mar. 2005.
3. S. M. Moghimi, A. C. Hunter, and J. C. Murray, "Nanomedicine: current status and future prospects," FASEB J., vol. 19, no. 3, pp. 311-330, Mar. 2005.

4. R. Singh and J. W. Lillard Jr., "Nanoparticle-based targeted drug delivery," Spec. Issue Struct. Biol., vol. 86, no. 3, pp. 215-223, Jun. 2009.
5. A. H. Faraji and P. Wipf, "Nanoparticles in cellular drug delivery," Bioorg. Med. Chem., vol. 17, no. 8, pp. 2950-2962, Apr. 2009.
6. G. M. Barratt, "Therapeutic applications of colloidal drug carriers," Pharm. Sci. Technol. Today, vol. 3, no. 5, pp. 163-171, May 2000.
7. D. Peer, J. M. Karp, S. Hong, O. C. Farokhzad, R. Margalit, and R. Langer, "Nanocarriers as an emerging platform for cancer therapy," Nat Nano, vol. 2, no. 12, pp. 751-760, Dec. 2007.
8. Z. Cheng, A. Al Zaki, J. Z. Hui, V. R. Muzykantov, and A. Tsourkas, "Multifunctional Nanoparticles: Cost Versus Benefit of Adding Targeting and Imaging Capabilities," Science, vol. 338, no. 6109, pp. 903-910, Nov. 2012.
9. O. C. Farokhzad and R. Langer, "Impact of Nanotechnology on Drug Delivery," ACS Nano, vol. 3, no. 1, pp. 16-20, Jan. 2009.
10. K. A. Dunphy Guzmán, M. R. Taylor, and J. F. Banfield, "Environmental Risks of Nanotechnology: National Nanotechnology Initiative Funding, 2000–2004," Environ. Sci. Technol., vol. 40, no. 5, pp. 1401-1407, Jan. 2006.
11. S. Bamrungsap, Z. Zhao, T. Chen, L. Wang, C. Li, T. Fu, and W. Tan, "Nanotechnology in therapeutics: a focus on nanoparticles as a drug delivery system," Nanomed., vol. 7, no. 8, pp. 1253-1271, Aug. 2012.
12. R. A. Petros and J. M. DeSimone, "Strategies in the design of nanoparticles for therapeutic applications," Nat Rev Drug Discov, vol. 9, no. 8, pp. 615-627, Aug. 2010.
13. V. R. Sinha and R. Kumria, "Polysaccharides in colon-specific drug delivery," Int. J. Pharm., vol. 224, no. 1-2, pp. 19-38, Aug. 2001.
14. R. Duncan, "The dawning era of polymer therapeutics," Nat Rev Drug Discov, vol. 2, no. 5, pp. 347-360, May 2003.
15. J. H. Park, S. Lee, J.-H. Kim, K. Park, K. Kim, and I. C. Kwon, "Polymeric nanomedicine for cancer therapy," Prog. Polym. Sci., vol. 33, no. 1, pp. 113-137, Jan. 2008.
16. H. Takeuchi, H. Yamamoto, and Y. Kawashima, "Mucoadhesive nanoparticulate systems for peptide drug delivery," Nanoparticulate Syst. Improv. Drug Deliv., vol. 47, no. 1, pp. 39-54, Mar. 2001.
17. V. M. Platt and F. C. Szoka, "Anticancer Therapeutics: Targeting Macromolecules and Nanocarriers to Hyaluronan or CD44, a Hyaluronan Receptor," Mol. Pharm., vol. 5, no. 4, pp. 474-486, Jun. 2008.

18. C. Lemarchand, R. Gref, and P. Couvreur, "Polysaccharide-decorated nanoparticles," Int. Assoc. Pharm. Technol. APV, vol. 58, no. 2, pp. 327-341, Sep. 2004.
19. A. D. Baldwin and K. L. Kiick, "Polysaccharide-modified synthetic polymeric biomaterials," Pept. Sci., vol. 94, no. 1, pp. 128-140, 2010.
20. Z. Liu, Y. Jiao, Y. Wang, C. Zhou, and Z. Zhang, "Polysaccharides-based nanoparticles as drug delivery systems," 2008 Ed. Collect., vol. 60, no. 15, pp. 1650-1662, Dec. 2008.
21. S. Mitra, U. Gaur, P. Ghosh, and A. Maitra, "Tumour targeted delivery of encapsulated dextran–doxorubicin conjugate using chitosan nanoparticles as carrier," J. Controlled Release, vol. 74, no. 1-3, pp. 317-323, Jul. 2001.
22. M. Bodnar, J. F. Hartmann, and J. Borbely, "Preparation and Characterization of Chitosan-Based Nanoparticles," Biomacromolecules, vol. 6, no. 5, pp. 2521-2527, Sep. 2005.
23. S. A. Agnihotri, N. N. Mallikarjuna, and T. M. Aminabhavi, "Recent advances on chitosan-based micro- and nanoparticles in drug delivery," J. Controlled Release, vol. 100, no. 1, pp. 5-28, Nov. 2004.
24. Y. S. Leong and F. Candau, "Inverse microemulsion polymerization," J. Phys. Chem., vol. 86, no. 13, pp. 2269-2271, Jun. 1982.
25. S. A. Agnihotri, N. N. Mallikarjuna, and T. M. Aminabhavi, "Recent advances on chitosan-based micro- and nanoparticles in drug delivery," J. Controlled Release, vol. 100, no. 1, pp. 5-28, Nov. 2004.
26. Q. Gan and T. Wang, "Chitosan nanoparticle as protein delivery carrier--systematic examination of fabrication conditions for efficient loading and release," Colloids Surf. B Biointerfaces, vol. 59, no. 1, pp. 24-34, Sep. 2007.
27. Y.-H. Lin, K. Sonaje, K. M. Lin, J.-H. Juang, F.-L. Mi, H.-W. Yang, and H.-W. Sung, "Multi-ion-crosslinked nanoparticles with pH-responsive characteristics for oral delivery of protein drugs," J. Control. Release Off. J. Control. Release Soc., vol. 132, no. 2, pp. 141-149, Dec. 2008.
28. N. Csaba, M. Köping-Höggård, E. Fernandez-Megia, R. Novoa-Carballal, R. Riguera, and M. J. Alonso, "Ionically crosslinked chitosan nanoparticles as gene delivery systems: effect of PEGylation degree on *in vitro* and *in vivo* gene transfer," J. Biomed. Nanotechnol., vol. 5, no. 2, pp. 162-171, Apr. 2009.
29. N. Csaba, M. Köping-Höggård, and M. J. Alonso, "Ionically crosslinked chitosan/tripolyphosphate nanoparticles for oligonucleotide and plasmid DNA delivery," Int. J. Pharm., vol. 382, no. 1-2, pp. 205-214, Dec. 2009.
30. K. A. Howard, U. L. Rahbek, X. Liu, C. K. Damgaard, S. Z. Glud, M. Ø. Andersen, M. B. Hovgaard, A. Schmitz, J. R. Nyengaard, F. Besenbacher, and J. Kjems, "RNA Interference *in vitro* and *in vivo* Using a

Chitosan/siRNA Nanoparticle System," Mol. Ther., vol. 14, no. 4, pp. 476-484, Oct. 2006.

31. K. A. Howard, S. R. Paludan, M. A. Behlke, F. Besenbacher, B. Deleuran, and J. Kjems, "Chitosan/siRNA Nanoparticle–mediated TNF-α Knockdown in Peritoneal Macrophages for Anti-inflammatory Treatment in a Murine Arthritis Model," Mol. Ther., vol. 17, no. 1, pp. 162-168, Sep. 2008.
32. H.-Q. Mao, K. Roy, V. L. Troung-Le, K. A. Janes, K. Y. Lin, Y. Wang, J. T. August, and K. W. Leong, "Chitosan-DNA nanoparticles as gene carriers: synthesis, characterization and transfection efficiency," J. Controlled Release, vol. 70, no. 3, pp. 399-421, Feb. 2001.
33. S. E. Kim, J. H. Park, Y. W. Cho, H. Chung, S. Y. Jeong, E. B. Lee, and I. C. Kwon, "Porous chitosan scaffold containing microspheres loaded with transforming growth factor-β1: Implications for cartilage tissue engineering," J. Controlled Release, vol. 91, no. 3, pp. 365-374, Sep. 2003.
34. J. P. Quiñones, Y. C. García, H. Curiel, and C. P. Covas, "Microspheres of chitosan for controlled delivery of brassinosteroids with biological activity as agrochemicals," Carbohydr. Polym., vol. 80, no. 3, pp. 915-921, May 2010.
35. Y. Pan, Y. Li, H. Zhao, J. Zheng, H. Xu, G. Wei, J. Hao, and F. Cui, "Bioadhesive polysaccharide in protein delivery system: chitosan nanoparticles improve the intestinal absorption of insulin *in vivo*," Int. J. Pharm., vol. 249, no. 1-2, pp. 139-147, Dec. 2002.
36. J.-O. You and C.-A. Peng, "Calcium-Alginate Nanoparticles Formed by Reverse Microemulsion as Gene Carriers," Macromol. Symp., vol. 219, no. 1, pp. 147-153, 2005.
37. A. H. E. Machado, D. Lundberg, A. J. Ribeiro, F. J. Veiga, B. Lindman, M. G. Miguel, and U. Olsson, "Preparation of Calcium Alginate Nanoparticles Using Water-in-Oil (W/O) Nanoemulsions," Langmuir, vol. 28, no. 9, pp. 4131-4141, Mar. 2012.
38. P. Calvo, C. Remuñán-López, J. L. Vila-Jato, and M. J. Alonso, "Novel hydrophilic chitosan-polyethylene oxide nanoparticles as protein carriers," J. Appl. Polym. Sci., vol. 63, no. 1, pp. 125-132, 1997.
39. K. A. Janes, P. Calvo, and M. J. Alonso, "Polysaccharide colloidal particles as delivery systems for macromolecules," Adv. Drug Deliv. Rev., vol. 47, no. 1, pp. 83-97, Mar. 2001.
40. A. Martínez, M. Benito-Miguel, I. Iglesias, J. M. Teijón, and M. D. Blanco, "Tamoxifen-loaded thiolated alginate-albumin nanoparticles as antitumoral drug delivery systems," J. Biomed. Mater. Res. A, vol. 100, no. 6, pp. 1467-1476, Jun. 2012.

41. R. Bodmeier, H. G. Chen, and O. Paeratakul, “A novel approach to the oral delivery of micro- or nanoparticles,” Pharm. Res., vol. 6, no. 5, pp. 413-417, May 1989.
42. Z. Ahmad, A. Zahoor, S. Sharma, and G. K. Khuller, “Inhalable alginate nanoparticles as antitubercular drug carriers against experimental tuberculosis,” Int. J. Antimicrob. Agents, vol. 26, no. 4, pp. 298-303, Oct. 2005.
43. M. Rajaonarivony, C. Vauthier, G. Couarraze, F. Puisieux, and P. Couvreur, “Development of a new drug carrier made from alginate,” J. Pharm. Sci., vol. 82, no. 9, pp. 912-917, 1993.
44. V. M. Platt and F. C. Szoka, “Anticancer Therapeutics: Targeting Macromolecules and Nanocarriers to Hyaluronan or CD44, a Hyaluronan Receptor,” Mol. Pharm., vol. 5, no. 4, pp. 474-486, Aug. 2008.
45. K. Y. Choi, G. Saravanakumar, J. H. Park, and K. Park, “Hyaluronic acid-based nanocarriers for intracellular targeting: Interfacial interactions with proteins in cancer,” Colloids Surf. B Biointerfaces, vol. 99, pp. 82-94, Nov. 2012.
46. B. P. Toole, “Hyaluronan: from extracellular glue to pericellular cue,” Nat. Rev. Cancer, vol. 4, no. 7, pp. 528-539, Jul. 2004.
47. K. Y. Choi, H. Chung, K. H. Min, H. Y. Yoon, K. Kim, J. H. Park, I. C. Kwon, and S. Y. Jeong, “Self-assembled hyaluronic acid nanoparticles for active tumor targeting,” Biomaterials, vol. 31, no. 1, pp. 106-114, Jan. 2010.
48. H. S. Han, J. Lee, H. R. Kim, S. Y. Chae, M. Kim, G. Saravanakumar, H. Y. Yoon, D. G. You, H. Ko, K. Kim, I. C. Kwon, J. C. Park, and J. H. Park, “Robust PEGylated hyaluronic acid nanoparticles as the carrier of doxorubicin: Mineralization and its effect on tumor targetability *in vivo*,” J. Controlled Release, vol. 168, no. 2, pp. 105–114, Jun. 2013.
49. S. Mizrahy and D. Peer, “Polysaccharides as building blocks for nanotherapeutics,” Chem. Soc. Rev., vol. 41, no. 7, pp. 2623–2640, 2012.
50. T. Ouchi, H. Nishizawa, and Y. Ohya, “Aggregation phenomenon of PEG-grafted chitosan in aqueous solution,” Polymer, vol. 39, no. 21, pp. 5171-5175, Oct. 1998.
51. R. Yoksan, M. Matsusaki, M. Akashi, and S. Chirachanchai, “Controlled hydrophobic/hydrophilic chitosan: colloidal phenomena and nanosphere formation,” Colloid Polym. Sci., vol. 282, no. 4, pp. 337–342, Feb. 2004.
52. P. Opanasopit, T. Ngawhirunpat, T. Rojanarata, C. Choochottiros, and S. Chirachanchai, “Camptothecin-incorporating N-phthaloylchitosan-g-mPEG self-assembly micellar system: Effect of degree of deacetylation,” Colloids Surf. B Biointerfaces, vol. 60, no. 1, pp. 117-124, Oct. 2007.

53. Y.-I. Jeong, S.-H. Kim, T.-Y. Jung, I.-Y. Kim, S.-S. Kang, Y.-H. Jin, H.-H. Ryu, H.-S. Sun, S. Jin, K.-K. Kim, K.-Y. Ahn, and S. Jung, "Polyion complex micelles composed of all-trans retinoic acid and poly (ethylene glycol)-grafted-chitosan," J. Pharm. Sci., vol. 95, no. 11, pp. 2348-2360, 2006.
54. X. Yang, Q. Zhang, Y. Wang, H. Chen, H. Zhang, F. Gao, and L. Liu, "Self-aggregated nanoparticles from methoxy poly(ethylene glycol)-modified chitosan: Synthesis; characterization; aggregation and methotrexate release *in vitro*," Colloids Surf. B Biointerfaces, vol. 61, no. 2, pp. 125-131, Feb. 2008.
55. H. Yu, W. Wang, X. Chen, C. Deng, and X. Jing, "Synthesis and characterization of the biodegradable polycaprolactone-graft-chitosan amphiphilic copolymers," Biopolymers, vol. 83, no. 3, pp. 233-242, 2006.
56. R. Gref, J. Rodrigues, and P. Couvreur, "Polysaccharides Grafted with Polyesters: Novel Amphiphilic Copolymers for Biomedical Applications," Macromolecules, vol. 35, no. 27, pp. 9861-9867, Dec. 2002.
57. C. Lemarchand, P. Couvreur, M. Besnard, D. Costantini, and R. Gref, "Novel Polyester-Polysaccharide Nanoparticles," Pharm. Res., vol. 20, no. 8, pp. 1284–1292, Aug. 2003.
58. W. Zhang, K. Gilstrap, L. Wu, R. B. K. C., M. A. Moss, Q. Wang, X. Lu, and X. He, "Synthesis and Characterization of Thermally Responsive Pluronic F127–Chitosan Nanocapsules for Controlled Release and Intracellular Delivery of Small Molecules," ACS Nano, vol. 4, no. 11, pp. 6747-6759, Nov. 2010.
59. J. S. Choi and H. S. Yoo, "Chitosan/Pluronic Hydrogel Containing bFGF/Heparin for Encapsulation of Human Dermal Fibroblasts," J. Biomater. Sci. Polym. Ed., vol. 24, no. 2, pp. 210-223, May 2012.
60. K. M. Park, S. Y. Lee, Y. K. Joung, J. S. Na, M. C. Lee, and K. D. Park, "Thermosensitive chitosan–Pluronic hydrogel as an injectable cell delivery carrier for cartilage regeneration," Acta Biomater., vol. 5, no. 6, pp. 1956-1965, Jul. 2009.
61. K. S. Oh, R. S. Kim, J. Lee, D. Kim, S. H. Cho, and S. H. Yuk, "Gold/chitosan/pluronic composite nanoparticles for drug delivery," J. Appl. Polym. Sci., vol. 108, no. 5, pp. 3239-3244, 2008.
62. Dai Hai Nguyen, Jong Hoon Choi, Yoon Ki Joung, and Ki Dong Park, "Disulfide-crosslinked heparin-pluronic nanogels as a redox-sensitive nanocarrier for intracellular protein delivery," J. Bioact. Compat. Polym., vol. 26, no. 3, pp. 287-300, May 2011.
63. S. H. Choi, J.-H. Lee, S.-M. Choi, and T. G. Park, "Thermally Reversible Pluronic/Heparin Nanocapsules Exhibiting 1000-Fold Volume Transition," Langmuir, vol. 22, no. 4, pp. 1758-1762, Jan. 2006.

64. L. Choisnard, A. Gèze, J.-L. Putaux, Y.-S. Wong, and D. Wouessidjewe, "Nanoparticles of β-Cyclodextrin Esters Obtained by Self-Assembling of Biotransesterified β-Cyclodextrins," Biomacromolecules, vol. 7, no. 2, pp. 515-520, Jan. 2006.

65. X.-G. Chen, C. M. Lee, and H.-J. Park, "O/W Emulsification for the Self-Aggregation and Nanoparticle Formation of Linoleic Acid Modified Chitosan in the Aqueous System," J. Agric. Food Chem., vol. 51, no. 10, pp. 3135-3139, Apr. 2003.

66. C.-G. Liu, K. G. H. Desai, X.-G. Chen, and H.-J. Park, "Linolenic Acid-Modified Chitosan for Formation of Self-Assembled Nanoparticles," J. Agric. Food Chem., vol. 53, no. 2, pp. 437-441, Dec. 2004.

67. C.-G. Liu, K. G. H. Desai, X.-G. Chen, and H.-J. Park, "Preparation and Characterization of Nanoparticles Containing Trypsin Based on Hydrophobically Modified Chitosan," J. Agric. Food Chem., vol. 53, no. 5, pp. 1728-1733, Feb. 2005.

68. F.-Q. Hu, G.-F. Ren, H. Yuan, Y.-Z. Du, and S. Zeng, "Shell cross-linked stearic acid grafted chitosan oligosaccharide self-aggregated micelles for controlled release of paclitaxel," Colloids Surf. B Biointerfaces, vol. 50, no. 2, pp. 97-103, Jul. 2006.

69. G.-B. Jiang, D. Quan, K. Liao, and H. Wang, "Novel Polymer Micelles Prepared from Chitosan Grafted Hydrophobic Palmitoyl Groups for Drug Delivery," Mol. Pharm., vol. 3, no. 2, pp. 152-160, Jan. 2006.

70. J. Zhang, X. G. Chen, Y. Y. Li, and C. S. Liu, "Self-assembled nanoparticles based on hydrophobically modified chitosan as carriers for doxorubicin," Nanomedicine Nanotechnol. Biol. Med., vol. 3, no. 4, pp. 258-265, Dec. 2007.

71. I. Lalush, H. Bar, I. Zakaria, S. Eichler, and E. Shimoni, "Utilization of Amylose–Lipid Complexes as Molecular Nanocapsules for Conjugated Linoleic Acid," Biomacromolecules, vol. 6, no. 1, pp. 121-130, Nov. 2004.

72. Y.-S. Wang, L.-R. Liu, Q. Jiang, and Q.-Q. Zhang, "Self-aggregated nanoparticles of cholesterol-modified chitosan conjugate as a novel carrier of epirubicin," Eur. Polym. J., vol. 43, no. 1, pp. 43-51, Jan. 2007.

73. K. Akiyoshi, S. Kobayashi, S. Shichibe, D. Mix, M. Baudys, S. Wan Kim, and J. Sunamoto, "Self-assembled hydrogel nanoparticle of cholesterol-bearing pullulan as a carrier of protein drugs: Complexation and stabilization of insulin," J. Controlled Release, vol. 54, no. 3, pp. 313-320, Aug. 1998.

74. K. Ikeda, T. Okada, S. Sawada, K. Akiyoshi, and K. Matsuzaki, "Inhibition of the formation of amyloid β-protein fibrils using biocompatible nanogels as artificial chaperones," FEBS Lett., vol. 580, no. 28–29, pp. 6587-6595, Dec. 2006.

75. K. Akiyoshi, E.-C. Kang, S. Kurumada, J. Sunamoto, T. Principi, and F. M. Winnik, "Controlled Association of Amphiphilic Polymers in Water: Thermosensitive Nanoparticles Formed by Self-Assembly of Hydrophobically Modified Pullulans and Poly(N-isopropylacrylamides)," Macromolecules, vol. 33, no. 9, pp. 3244-3249, Apr. 2000.
76. K. Y. Lee, J.-H. Kim, I. C. Kwon, and S. Y. Jeong, "Self-aggregates of deoxycholic acid-modified chitosan as a novel carrier of adriamycin," Colloid Polym. Sci., vol. 278, no. 12, pp. 1216-1219, Dec. 2000.
77. K. Lee, I. Kwon, Y.-H. Kim, W. Jo, and S. Jeong, "Preparation of chitosan self-aggregates as a gene delivery system," J. Controlled Release, vol. 51, no. 2-3, pp. 213-220, Feb. 1998.
78. S. Y. Chae, S. Son, M. Lee, M.-K. Jang, and J.-W. Nah, "Deoxycholic acid-conjugated chitosan oligosaccharide nanoparticles for efficient gene carrier," Proc. Twelfth Int. Symp. Recent Adv. Drug Deliv. Syst., vol. 109, no. 1-3, pp. 330-344, Dec. 2005.
79. K. Kim, S. Kwon, J. H. Park, H. Chung, S. Y. Jeong, I. C. Kwon, and I.-S. Kim, "Physicochemical Characterizations of Self-Assembled Nanoparticles of Glycol Chitosan−Deoxycholic Acid Conjugates," Biomacromolecules, vol. 6, no. 2, pp. 1154-1158, Jan. 2005.
80. J. H. Park, S. Kwon, J.-O. Nam, R.-W. Park, H. Chung, S. B. Seo, I.-S. Kim, I. C. Kwon, and S. Y. Jeong, "Self-assembled nanoparticles based on glycol chitosan bearing 5β-cholanic acid for RGD peptide delivery," J. Controlled Release, vol. 95, no. 3, pp. 579-588, Mar. 2004.
81. H. Sang Yoo, J. Eun Lee, H. Chung, I. Chan Kwon, and S. Young Jeong, "Self-assembled nanoparticles containing hydrophobically modified glycol chitosan for gene delivery," J. Controlled Release, vol. 103, no. 1, pp. 235-243, Mar. 2005.
82. J.-H. Kim, Y.-S. Kim, S. Kim, J. H. Park, K. Kim, K. Choi, H. Chung, S. Y. Jeong, R.-W. Park, I.-S. Kim, and I. C. Kwon, "Hydrophobically modified glycol chitosan nanoparticles as carriers for paclitaxel," J. Controlled Release, vol. 111, no. 1-2, pp. 228-234, Mar. 2006.
83. K. Kim, J. H. Kim, H. Park, Y.-S. Kim, K. Park, H. Nam, S. Lee, J. H. Park, R.-W. Park, I.-S. Kim, K. Choi, S. Y. Kim, K. Park, and I. C. Kwon, "Tumor-homing multifunctional nanoparticles for cancer theragnosis: Simultaneous diagnosis, drug delivery, and therapeutic monitoring," Nanomedicine Drug Deliv. NanoDDS09, vol. 146, no. 2, pp. 219-227, Sep. 2010.
84. J. S. Park, T. H. Han, K. Y. Lee, S. S. Han, J. J. Hwang, D. H. Moon, S. Y. Kim, and Y. W. Cho, "N-acetyl histidine-conjugated glycol chitosan self-assembled nanoparticles for intracytoplasmic delivery of drugs:

Endocytosis, exocytosis and drug release," J. Controlled Release, vol. 115, no. 1, pp. 37-45, Sep. 2006.

85. K. Park, K. Kim, I. C. Kwon, S. K. Kim, S. Lee, D. Y. Lee, and Y. Byun, "Preparation and Characterization of Self-Assembled Nanoparticles of Heparin-Deoxycholic Acid Conjugates," Langmuir, vol. 20, no. 26, pp. 11726-11731, Nov. 2004.
86. K. Y. Choi, K. H. Min, J. H. Na, K. Choi, K. Kim, J. H. Park, I. C. Kwon, and S. Y. Jeong, "Self-assembled hyaluronic acid nanoparticles as a potential drug carrier for cancer therapy: synthesis, characterization, and *in vivo* biodistribution," J. Mater. Chem., vol. 19, no. 24, pp. 4102-4107, 2009.
87. G. Saravanakumar, K. Y. Choi, H. Y. Yoon, K. Kim, J. H. Park, I. C. Kwon, and K. Park, "Hydrotropic hyaluronic acid conjugates: Synthesis, characterization, and implications as a carrier of paclitaxel," Int. J. Pharm., vol. 394, no. 1-2, pp. 154-161, Jul. 2010.
88. Y. J. Son, J.-S. Jang, Y. W. Cho, H. Chung, R.-W. Park, I. C. Kwon, I.-S. Kim, J. Y. Park, S. B. Seo, C. R. Park, and S. Y. Jeong, "Biodistribution and anti-tumor efficacy of doxorubicin loaded glycol-chitosan nanoaggregates by EPR effect," Proc. Second Int. Symp. Tumor Target. Deliv. Syst., vol. 91, no. 1-2, pp. 135-145, Aug. 2003.
89. J. Hyung Park, S. Kwon, M. Lee, H. Chung, J.-H. Kim, Y.-S. Kim, R.-W. Park, I.-S. Kim, S. Bong Seo, I. C. Kwon, and S. Young Jeong, "Self-assembled nanoparticles based on glycol chitosan bearing hydrophobic moieties as carriers for doxorubicin: *In vivo* biodistribution and anti-tumor activity," Biomaterials, vol. 27, no. 1, pp. 119–126, Jan. 2006.
90. Y. W. Cho, S. A. Park, T. H. Han, D. H. Son, J. S. Park, S. J. Oh, D. H. Moon, K.-J. Cho, C.-H. Ahn, Y. Byun, I.-S. Kim, I. C. Kwon, and S. Y. Kim, "*In vivo* tumor targeting and radionuclide imaging with self-assembled nanoparticles: Mechanisms, key factors, and their implications," Biomaterials, vol. 28, no. 6, pp. 1236-1247, Feb. 2007.
91. Z. Aiping, C. Tian, Y. Lanhua, W. Hao, and L. Ping, "Synthesis and characterization of N-succinyl-chitosan and its self-assembly of nanospheres," Carbohydr. Polym., vol. 66, no. 2, pp. 274-279, Oct. 2006.
92. K. Na, T. Bum Lee, K.-H. Park, E.-K. Shin, Y.-B. Lee, and H.-K. Choi, "Self-assembled nanoparticles of hydrophobically-modified polysaccharide bearing vitamin H as a targeted anti-cancer drug delivery system," Eur. J. Pharm. Sci., vol. 18, no. 2, pp. 165-173, Feb. 2003.
93. C. Passirani, G. Barratt, J.-P. Devissaguet, and D. Labarre, "Interactions of nanoparticles bearing heparin or dextran covalently bound to poly(methyl methacrylate) with the complement system," Life Sci., vol. 62, no. 8, pp. 775-785, Jan. 1998.

94. C. Passirani, G. Barratt, J.-P. Devissaguet, and D. Labarre, "Long-Circulating Nanopartides Bearing Heparin or Dextran Covalently Bound to Poly(Methyl Methacrylate)," Pharm. Res., vol. 15, no. 7, pp. 1046-1050, Jul. 1998.

95. I. Bertholon, C. Vauthier, and D. Labarre, "Complement Activation by Core–Shell Poly(isobutylcyanoacrylate)–Polysaccharide Nanoparticles: Influences of Surface Morphology, Length, and Type of Polysaccharide," Pharm. Res., vol. 23, no. 6, pp. 1313-1323, Jun. 2006.

96. S. C. Yang, H. X. Ge, Y. Hu, X. Q. Jiang, and C. Z. Yang, "Formation of positively charged poly(butyl cyanoacrylate) nanoparticles stabilized with chitosan," Colloid Polym. Sci., vol. 278, no. 4, pp. 285-292, Apr. 2000.

97. C. Chauvierre, D. Labarre, P. Couvreur, and C. Vauthier, "Novel Polysaccharide-Decorated Poly(Isobutyl Cyanoacrylate) Nanoparticles," Pharm. Res., vol. 20, no. 11, pp. 1786-1793, Nov. 2003.

98. C. Chauvierre, C. Vauthier, D. Labarre, and H. Hommel, "Evaluation of the surface properties of dextran-coated poly(isobutylcyanoacrylate) nanoparticles by spin-labelling coupled with electron resonance spectroscopy," Colloid Polym. Sci., vol. 282, no. 9, pp. 1016-1025, Jul. 2004.

99. I. Bravo-Osuna, G. Millotti, C. Vauthier, and G. Ponchel, "*In vitro* evaluation of calcium binding capacity of chitosan and thiolated chitosan poly(isobutyl cyanoacrylate) core–shell nanoparticles," Int. J. Pharm., vol. 338, no. 1–2, pp. 284–290, Jun. 2007.

100. D. Labarre, C. Vauthier, C. Chauvierre, B. Petri, R. Müller, and M. M. Chehimi, "Interactions of blood proteins with poly(isobutylcyanoacrylate) nanoparticles decorated with a polysaccharidic brush," Biomaterials, vol. 26, no. 24, pp. 5075–5084, Aug. 2005.

101. C. Chauvierre, M. C. Marden, C. Vauthier, D. Labarre, P. Couvreur, and L. Leclerc, "Heparin coated poly(alkylcyanoacrylate) nanoparticles coupled to hemoglobin: a new oxygen carrier," Biomaterials, vol. 25, no. 15, pp. 3081-3086, Jul. 2004.

102. R. P. Brinkhuis, F. P. J. T. Rutjes, and J. C. M. van Hest, "Polymeric vesicles in biomedical applications," Polym. Chem., vol. 2, no. 7, pp. 1449-1462, 2011.

103. C. Schatz, S. Louguet, J.-F. Le Meins, and S. Lecommandoux, "Polysaccharide-block-polypeptide Copolymer Vesicles: Towards Synthetic Viral Capsids," Angew. Chem. Int. Ed., vol. 48, no. 14, pp. 2572-2575, 2009.

104. K. K. Upadhyay, A. N. Bhatt, A. K. Mishra, B. S. Dwarakanath, S. Jain, C. Schatz, J.-F. Le Meins, A. Farooque, G. Chandraiah, A. K. Jain, A. Misra, and S. Lecommandoux, "The intracellular drug delivery and anti tumor

activity of doxorubicin loaded poly(γ-benzyl l-glutamate)-b-hyaluronan polymersomes," Biomaterials, vol. 31, no. 10, pp. 2882-2892, Apr. 2010.

105. I. Otsuka, C. Travelet, S. Halila, S. Fort, I. Pignot-Paintrand, A. Narumi, and R. Borsali, "Thermoresponsive Self-Assemblies of Cyclic and Branched Oligosaccharide-block-poly(N-isopropylacrylamide) Diblock Copolymers into Nanoparticles," Biomacromolecules, vol. 13, no. 5, pp. 1458-1465, Mar. 2012.

106. M. Felici, M. Marzá-Pérez, N. S. Hatzakis, R. J. M. Nolte, and M. C. Feiters, "β-Cyclodextrin-Appended Giant Amphiphile: Aggregation to Vesicle Polymersomes and Immobilisation of Enzymes," Chem. – Eur. J., vol. 14, no. 32, pp. 9914-9920, 2008.

107. Z. Cui and R. J. Mumper, "Chitosan-based nanoparticles for topical genetic immunization," J. Controlled Release, vol. 75, no. 3, pp. 409-419, Aug. 2001.

108. Y. Chen, V. Mohanraj, and J. Parkin, "Chitosan-dextran sulfate nanoparticles for delivery of an anti-angiogenesis peptide," Lett. Pept. Sci., vol. 10, no. 5-6, pp. 621-629, Sep. 2003.

109. W. Tiyaboonchai and N. Limpeanchob, "Formulation and characterization of amphotericin B–chitosan–dextran sulfate nanoparticles," Int. J. Pharm., vol. 329, no. 1-2, pp. 142-149, Feb. 2007.

110. M. Huang, S. N. Vitharana, L. J. Peek, T. Coop, and C. Berkland, "Polyelectrolyte Complexes Stabilize and Controllably Release Vascular Endothelial Growth Factor," Biomacromolecules, vol. 8, no. 5, pp. 1607-1614, Apr. 2007.

111. B. Sarmento, S. Martins, A. Ribeiro, F. Veiga, R. Neufeld, and D. Ferreira, "Development and Comparison of Different Nanoparticulate Polyelectrolyte Complexes as Insulin Carriers," Int. J. Pept. Res. Ther., vol. 12, no. 2, pp. 131-138, Jun. 2006.

112. B. Sarmento, D. Ferreira, F. Veiga, and A. Ribeiro, "Characterization of insulin-loaded alginate nanoparticles produced by ionotropic pre-gelation through DSC and FTIR studies," Carbohydr. Polym., vol. 66, no. 1, pp. 1-7, Oct. 2006.

113. S. Cafaggi, E. Russo, R. Stefani, R. Leardi, G. Caviglioli, B. Parodi, G. Bignardi, D. De Totero, C. Aiello, and M. Viale, "Preparation and evaluation of nanoparticles made of chitosan or N-trimethyl chitosan and a cisplatin–alginate complex," Fourth Int. Nanomedicine Drug Deliv. Symp., vol. 121, no. 1-2, pp. 110-123, Aug. 2007.

114. T. Li, X.-W. Shi, Y.-M. Du, and Y.-F. Tang, "Quaternized chitosan/alginate nanoparticles for protein delivery," J. Biomed. Mater. Res. A, vol. 83A, no. 2, pp. 383-390, 2007.

115. M. Alonso-Sande, M. Cuña, C. Remuñán-López, D. Teijeiro-Osorio, J. L. Alonso-Lebrero, and M. J. Alonso, "Formation of New Glucomannan–Chitosan Nanoparticles and Study of Their Ability To Associate and Deliver Proteins," Macromolecules, vol. 39, no. 12, pp. 4152-4158, May 2006.

116. J. Du, S. Zhang, R. Sun, L.-F. Zhang, C.-D. Xiong, and Y.-X. Peng, "Novel polyelectrolyte carboxymethyl konjac glucomannan–chitosan nanoparticles for drug delivery. II. Release of albumin *in vitro*," J. Biomed. Mater. Res. B Appl. Biomater., vol. 72B, no. 2, pp. 299-304, 2005.

117. Z. Liu, Y. Jiao, F. Liu, and Z. Zhang, "Heparin/chitosan nanoparticle carriers prepared by polyelectrolyte complexation," J. Biomed. Mater. Res. A, vol. 83A, no. 3, pp. 806-812, 2007.

118. S. Mao, U. Bakowsky, A. Jintapattanakit, and T. Kissel, "Self-assembled polyelectrolyte nanocomplexes between chitosan derivatives and insulin," J. Pharm. Sci., vol. 95, no. 5, pp. 1035–1048, 2006.

119. T. Ishii, Y. Okahata, and T. Sato, "Mechanism of cell transfection with plasmid/chitosan complexes," Biochim. Biophys. Acta BBA - Biomembr., vol. 1514, no. 1, pp. 51–64, Sep. 2001.

120. H. de Martimprey, C. Vauthier, C. Malvy, and P. Couvreur, "Polymer nanocarriers for the delivery of small fragments of nucleic acids: Oligonucleotides and siRNA," Eng. Polym. Control. Drug Deliv. Target., vol. 71, no. 3, pp. 490-504, Mar. 2009.

121. M. George and T. E. Abraham, "Polyionic hydrocolloids for the intestinal delivery of protein drugs: Alginate and chitosan — a review," J. Controlled Release, vol. 114, no. 1, pp. 1-14, Aug. 2006.

122. Y.-J. Kim, S. Y. Chae, C.-H. Jin, M. Sivasubramanian, S. Son, K. Y. Choi, D.-G. Jo, K. Kim, I. Chan Kwon, K. C. Lee, and J. H. Park, "Ionic complex systems based on hyaluronic acid and PEGylated TNF-related apoptosis-inducing ligand for treatment of rheumatoid arthritis," Biomaterials, vol. 31, no. 34, pp. 9057-9064, Dec. 2010.

123. N. Davidenko, M. D. Blanco, C. Peniche, L. Becherán, S. Guerrero, and J. M. Teijón, "Effects of different parameters on the characteristics of chitosan–poly(acrylic acid) nanoparticles obtained by the method of coacervation," J. Appl. Polym. Sci., vol. 111, no. 5, pp. 2362-2371, 2009.

124. Q. Chen, Y. Hu, Y. Chen, X. Jiang, and Y. Yang, "Microstructure Formation and Property of Chitosan-Poly(acrylic acid) Nanoparticles Prepared by Macromolecular Complex," Macromol. Biosci., vol. 5, no. 10, pp. 993-1000, 2005.

125. P.-W. Lee, S.-F. Peng, C.-J. Su, F.-L. Mi, H.-L. Chen, M.-C. Wei, H.-J. Lin, and H.-W. Sung, "The use of biodegradable polymeric nanoparticles in

combination with a low-pressure gene gun for transdermal DNA delivery," Biomaterials, vol. 29, no. 6, pp. 742-751, Feb. 2008.

126. Y. Zheng, W. Yang, C. Wang, J. Hu, S. Fu, L. Dong, L. Wu, and X. Shen, "Nanoparticles based on the complex of chitosan and polyaspartic acid sodium salt: Preparation, characterization and the use for 5-fluorouracil delivery," Eur. J. Pharm. Biopharm., vol. 67, no. 3, pp. 621-631, Nov. 2007.

127. Y. Zheng, Y. Wu, W. Yang, C. Wang, S. Fu, and X. Shen, "Preparation, characterization, and drug release *in vitro* of chitosan-glycyrrhetic acid nanoparticles," J. Pharm. Sci., vol. 95, no. 1, pp. 181-191, 2006.

128. S. Sajeesh and C. P. Sharma, "Novel pH responsive polymethacrylic acid–chitosan–polyethylene glycol nanoparticles for oral peptide delivery," J. Biomed. Mater. Res. B Appl. Biomater., vol. 76B, no. 2, pp. 298–305, 2006.

129. B. G. De Geest, G. B. Sukhorukov, and H. Möhwald, "The pros and cons of polyelectrolyte capsules in drug delivery," Expert Opin. Drug Deliv., vol. 6, no. 6, pp. 613–624, Jun. 2009.

130. C. E. Mora-Huertas, H. Fessi, and A. Elaissari, "Polymer-based nanocapsules for drug delivery," Int. J. Pharm., vol. 385, no. 1–2, pp. 113-142, Jan. 2010.

131. T. Crouzier, T. Boudou, and C. Picart, "Polysaccharide-based polyelectrolyte multilayers," Curr. Opin. Colloid Interface Sci., vol. 15, no. 6, pp. 417-426, Dec. 2010.

132. L. J. De Cock, S. De Koker, B. G. De Geest, J. Grooten, C. Vervaet, J. P. Remon, G. B. Sukhorukov, and M. N. Antipina, "Polymeric Multilayer Capsules in Drug Delivery," Angew. Chem. Int. Ed., vol. 49, no. 39, pp. 6954-6973, 2010.

133. W. Wang, C. Luo, S. Shao, and S. Zhou, "Chitosan hollow nanospheres fabricated from biodegradable poly-d,l-lactide-poly(ethylene glycol) nanoparticle templates," Eur. J. Pharm. Biopharm., vol. 76, no. 3, pp. 376-383, Nov. 2010.

134. Y. Hu, Y. Ding, D. Ding, M. Sun, L. Zhang, X. Jiang, and C. Yang, "Hollow Chitosan/Poly(acrylic acid) Nanospheres as Drug Carriers," Biomacromolecules, vol. 8, no. 4, pp. 1069-1076, Feb. 2007.

135. Y. Yin, S. Xu, D. Chang, H. Zheng, J. Li, X. Liu, P. Xu, and F. Xiong, "One-pot synthesis of biopolymeric hollow nanospheres by photocrosslinking," Chem. Commun., vol. 46, no. 43, pp. 8222-8224, 2010.